Correspondence Analysis and Data Coding with Java and R

Chapman & Hall/CRC
Computer Science and Data Analysis Series

The interface between the computer and statistical sciences is increasing, as each discipline seeks to harness the power and resources of the other. This series aims to foster the integration between the computer sciences and statistical, numerical, and probabilistic methods by publishing a broad range of reference works, textbooks, and handbooks.

SERIES EDITORS
John Lafferty, Carnegie Mellon University
David Madigan, Rutgers University
Fionn Murtagh, Royal Holloway, University of London
Padhraic Smyth, University of California, Irvine

Proposals for the series should be sent directly to one of the series editors above, or submitted to:

Chapman & Hall/CRC
23-25 Blades Court
London SW15 2NU
UK

Published Titles

Bayesian Artificial Intelligence
Kevin B. Korb and Ann E. Nicholson

Pattern Recognition Algorithms for Data Mining
Sankar K. Pal and Pabitra Mitra

Exploratory Data Analysis with MATLAB®
Wendy L. Martinez and Angel R. Martinez

Clustering for Data Mining: A Data Recovery Approach
Boris Mirkin

Correspondence Analysis and Data Coding with Java and R
Fionn Murtagh

R Graphics
Paul Murrell

Computer Science and Data Analysis Series

Correspondence Analysis and Data Coding with Java and R

Fionn Murtagh

Foreword by J.-P. Benzécri

CRC Press
Taylor & Francis Group
Boca Raton London New York

CRC Press is an imprint of the
Taylor & Francis Group, an **informa** business
A CHAPMAN & HALL BOOK

CRC Press
Taylor & Francis Group
6000 Broken Sound Parkway NW, Suite 300
Boca Raton, FL 33487-2742

First issued in paperback 2019

ISBN-13: 978-1-58488-528-3 (hbk)
ISBN-13: 978-0-367-39273-4 (pbk)

Library of Congress Card Number 2005041353

Library of Congress Cataloging-in-Publication Data

Murtagh, Fionn.
    Correspondence analysis and data coding with Java and R / by Fionn Murtagh.
        p. cm.
    Includes bibliographical references and index.
    ISBN 1-58488-528-9
    1. Java (Computer program language) 2. R (Computer program language). 3. Correspondence analysis (Statistics) I. Title.

QA76.73.J38M877 2005
005.13'3--dc22

                                                    2005041353

Visit the Taylor & Francis Web site at
http://www.taylorandfrancis.com

and the CRC Press Web site at
http://www.crcpress.com

Preface

Some years ago, the correspondence analysis framework for analyzing data was very widespread, in particular in France. A very large number of students passed through the doctoral school of Jean-Paul Benzécri in Paris. Many books were available, most of them in French. Many contemporary prominent researchers passed through Jean-Paul Benzécri's graduate school. One objective of this book is to provide accessible inroads into this tradition.

A second objective is to show how and where the topicality and importance of correspondence analysis remain. We will exemplify this in areas such as the analysis of time-evolving data, and analysis of text. We will show how correspondence analysis remains unsurpassed as an analysis framework.

In this book, too, we give a very clear reason for the continuing importance of correspondence analysis. We will not spoil the reading of this book by stating straight away what this is. It is simply, but profoundly, that data coding is handled so well by the correspondence analysis framework. The correspondence analysis framework offers not so much a theory of data, nor a theory of data interpretation, as a philosophy of – a conceptual approach to – data, and data interpretation.

To support our work in writing this book we used software in the R statistical language, and in the Java language, both of which relied on earlier software of ours or of others, in other languages that are now less widely used. This software, together with data sets used in this work, are available at the book's web site:

www.correspondances.info

Note that *correspondances* is spelt as in French, with an "a."

Chapter 1 begins with a historical description of some important developments which have led to today's multivariate and multidimensional data analysis. The data analysis method known as principal components analysis is widely used in statistics, pattern recognition, signal processing, and many other areas, so we discuss in the form of a case study the application of it, principal components analysis, and correspondence analysis, to the same data. Then, in this first chapter, we begin the description of software, which will be continued, and illustrated with examples, in subsequent chapters.

Chapter 2 is a theoretical overview of the mathematics and the underlying algorithms of correspondence analysis, together with hierarchical clustering.

It can be read in its own right, or it can be dipped into as the need, or interest, arises.

Chapter 3 addresses an area of data analysis that has not been given due attention in the past, but which is handled by quite a formidable range of software tools and options in the correspondence analysis tradition: this area is that of data coding. Clearly, the results of an analysis depend crucially on the data that are analyzed. The very varied possibilities offered by correspondence analysis are surveyed in this chapter. At the end of this chapter, the Java software is introduced, for correspondence analysis, for clustering, and interpretation tools. This software is independent of the R code presented in other chapters.

Chapter 4 presents and discusses a range of case studies. The software used is discussed.

Chapter 5 firstly reviews the wealth of studies carried out on text by Jean-Paul Benzécri's research and teaching lab over a number of decades. Based on this rich tradition, we pose the question: can correspondence analysis allow understanding of textual content, in addition to textual form? We want to show the importance of correspondence analysis in artificial intelligence, in addition to stylometry and other fields. Through various examples we illustrate the textual analysis of content, as well as form. Software – for convenience in C – for analyzing text is available, and discussed. Then we move on to a large range of new studies, all with our aim of showing the relevance and utility of correspondence analysis in fields concerned with the analysis of textual information.

The first and foremost acknowledgement for this work is to Jean-Paul Benzécri. May he find an accurate albeit limited reflection here of his results and writings over many decades. Much inspiration and motivation came from discussions with T.K. Gopalan (Chennai), and material also for parts of chapters 2 and 3.

Early on, questions of motivation were raised by Jonathan Campbell (Derry), and were addressed in the work represented here. For extensive comments on an early draft of this book, leading to much reshaping, I am indebted to Josiane Mothe (Toulouse) and Boris Mirkin (London). Dimitri Zervas wrote the programs in C for text analysis. Some material here was presented in an early form at the IASC/IFCS (International Association for Statistical Computing – International Federation of Classification Societies) Joint International Summer School, JISS-2003, on Classification and Data Mining in Business, Industry and Applied Research – Methodological and Computational Issues, held at the Faculty of Psychology and Education Science, University of Lisbon, in July 2003, and organized by Helena Bacelar Nicolau. For early discussions on analysis of the Ross oil data, I am indebted to Pat Muldowney (Derry), which was in the context of a British Council project with Polish partners.

Fionn Murtagh, Royal Holloway, University of London

Avant-Propos

Quand, en 1954–55 je séjournais à Princeton, il n'y avait d'ordinateur (on disait: Computer), ni à l'Université, ni à l'Institute for Advanced Studies. Un étudiant pouvait consacrer une année à fabriquer un calculateur analogique, destiné à résoudre des problèmes techniques d'une catégorie particulière. Et il en était de même au Laboratoire de Physique de l'École Normale Supérieure.

Je pensais que les ordinateurs ne pouvaient être que des merveilles; mais que, bien peu y ayant accès, il était sage de recourir à des simplifications mathématiques radicales, afin de renfermer les calculs dans les limites du possible.

C'est ainsi que, de 1959 à 1960, étant au Groupe de Recherche Opérationnelle de la Marine Nationale, je ne me faisais pas de scrupule de réduire à une loi normale toute donnée multidimensionelle; collaborant parfois avec un camarade pour des simulations graphiques.

Pourtant, quand, sur le projet de la Traduction Automatique des Langues Naturelles, linguistique, logique et mathématique, entreprirent de collaborer en ayant l'ordinateur pour outil ..., il apparut que, dans la voie frayée par Louis Guttman † et Chikio Hayashi †, le principe d'équivalence distributionnelle, proposé par le linguiste Zelig Harris †, devait régler l'analyse des données statistiques.

Alors, en donnant forme géométrique à cette analyse, on aboutirait à la recherche des axes principaux d'inertie d'un nuage de points munis de masse; problème classique, en dimension 3, mais à traiter ici en une dimension, n, quelconque. Ce qui requiert, impérativement, des diagonalisations de matrices carrées $n \times n$, calcul infaisable sans une machine, dès que n dépasse 3 (ou 4 ...).

Vers 1963, diagonaliser une matrice 7×7, était, pour un laboratoire modestement équipé, une tâche considérable. Par la suite, la Classification Ascendante Hiérarchique demanda des calculs encore plus lourds que ceux des diagonalisations. Mais la puissance des machines croissant avec l'efficacité des algorithmes, notre carrière de statisticien se développa...; en mettant au service d'ambitions croissant sans cesse, des techniques dont le progrès défiait tous les rêves!

Vers 2000, sur un micro-ordinateur tel que ceux offerts à la clientèle des marchés, on peut, en quelques minutes, classer plusieurs milliers d'individus. Plus exactement, il faut quelques minutes pour les algorithmes de classification et d'analyse factorielle... Mais la conception des données, leur mise en forme, l'examen des résultats prennent non seulement des heures, mais des mois...

Il n'y a plus, à strictement parler, de problème de calcul; mais le problème même de l'Analyse des données subsiste; d'autant plus vaste que, le calcul ne mettant point de borne à la recherche, on n'a point d'excuse pour s'arrêter dans la collecte des données et la méditation. Relativement à 1960..., le rapport de difficulté, entre projets intellectuels et calculs, est inversé.

Il s'en faut de beaucoup que les principes qui nous paraissent s'imposer soient admis de tous.

Quant à la philosophie des nombres, la distinction entre qualitatif et quantitatif ne nous semble pas être toujours bien comprise. En bref, il ne faut pas dire:

- grandeur numérique continue $\approx$ donnée quantitative;

- grandeur à un nombre fini de modalités $\approx$ donnée qualitative;

car au niveau de l'individu statistique (e.g., le dossier d'un malade), une donnée numérique: l'âge, ou même: la pression artérielle ou la glycémie, n'est généralement pas à prendre avec toute sa précision, mais selon sa signification; et, de ce point de vue, il n'y a pas de différence de nature entre âge et profession.

Et surtout, pour comparer un individu à un autre, il faut considérer, non deux ensembles de données primaires, par exemple deux ensembles de cent nombres réels, un point de $\mathbb{R}^{100}$, à un autre point de $\mathbb{R}^{100}$, entre lesquels des ressemblances globales ne se voient pas, mais la synthèse de ces ensembles, aboutissant à quelques gradations, ou à des discontinuités, à des diagnostiques...

Quant au calcul, les algorithmes d'analyse factorielle (dont on a dit que le coût numérique est celui d'une diagonalisation de matrice) et de classification ascendante hiérarchique, jouant sur des données codées suivant le principe global d'équivalence distributionnelle (de profil), l'emportent en efficacité sur le jeu des contiguïtés entre individus, pris en compte par les algorithmes d'approximation stochastique, souvent effectués, aujourd'hui, suivant le schéma des réseaux de neurones.

Tel est le Monde vu par un statisticien géomètre après quarante ans de pratique.

Est-il permis d'assimiler le monde à ce qu'on a vu et vécu? Prétention commune, aussi condamnable que le refus de rêver – au moins (faute de mieux) – sur l'avenir!

Premièrement: la question de reconnaître l'ordre de ce qui est dans les éléments que saisissent les sens (ou les outils qui arment les sens) est peut-être celle-même de la Philosophie, dans toute sa noblesse. On a dit, en latin... que toute connaissance commence par ce qui est sensible dans les objets de la nature: "Omnis cognitio initium habet a naturalibus... vel: a sensibilibus". Au-delà de cette connaissance, il n'y a que le mystère; et la révélation même, donnée par Dieu, se médite à l'exemple de ce qu'on a pu connaître par le jeu naturel de la raison.

Il faut ici que le statisticien, le géomètre, le sociologue soient modestes! En cherchant ce qu'on a toujours dû chercher, chaque génération ne peut avoir fait plus que sa part: la question subsiste.

Deuxième: on voit sur l'exemple des mathématiques, que le calcul nouveau, dont la vitesse dépasse celle du calcul de notre génération dans un rapport aussi inimaginable aujourd'hui que ne l'était, il y a un demi-siècle, le rapport de la complexité des calculs que nous avons pu faire à celle des calculs antérieurs... on voit, dis-je, qu'il ne faut pas trop vite affirmer, méprisement, que la pensée ne peut que devenir paresseuse quand l'outil devient plus puissant.

D'une part, afin de résoudre des problèmes de calcul intégral, on a inventé des "fonctions spéciales"; et, chemin faisant, on a créé l'analyse des fonctions de variable complexe (ou, du moins, approfondi cette analyse). De même, pour l'intégration des équations aux dérivées partielles, laquelle demande: la théorie des espaces fonctionnels. Aujourd'hui, tous les calculs pratiques semblent être réduits au jeu banal des méthodes les plus simples, sur des réseaux de points arbitrairement denses...

En somme, le problème pratique provoque (ou, du moins, aiguillonne) le développement des idées théoriques; et le perfectionnement des outils rend paresseuse la spéculation théorique.

Cependant, le mouvement inverse existe aussi. On remarque des coïncidences; et on donne à ces coïncidences forme de lois, mises en circulation, avant que le développement d'idées théoriques appropriés permette de démontrer ces lois.

La théorie des fonctions analytiques doit beaucoup au désir de démontrer le grand théorème de Fermat: $x^n + y^n = z^n$ n'a pas de solution pour des entiers $n > 2$.

Or Fermat n'a pu conjecturer qu'après avoir calculé ... remarqué ... essayé de renverser sa remarque ou de la confirmer; et cela, avec les moyens de calcul de son temps.

Voici un exemple trouvé, par internet: le résumé d'un article de Théorie physique des hautes énergies:

hep-th/9811173 19 Nov 1998; @ http://xxx.lanl.gov/

résumé qui intéressera ceux mêmes qui ne sont pas mieux avertis que moi de la théorie des fonctions zeta géréralisées et de l'analyse des diagrammes de Feynman (de la Théorie Quantique des Champs) faite en dénombrant des noeuds.

Au prix de calculs formidables, on aboutit au résultat que des fractions, dont le numérateur, "a", peut être de l'ordre de un million, n'ont pas de numérateur, "b", supérieur à 9.

Avec, à la fin du résumé cette conclusion.

Nos résultats sont sûrs, numériquement; mais il semble bien difficile de les démontrer par l'analyse.

"Sûrs" étant à comprendre en termes statistiques: si l'on procède par tirage aléatoire, il n'y a pour ainsi dire pas de tels ensembles de fractions, etc. ..., présentant les propriétés arithmétiques qu'on a trouvées.

On sera peut-être encore plus surpris de trouver, dans ce même domaine de la Physique des hautes énergies, des conclusions telles que la suivante:

Le résultat de la présente étude ne fait aucun doûte... Cependant, pour décider de telle autre question..., il faudrait des calculs mille fois plus complexes; calculs présentement inabordables.

Mais comme, répétons-le, la rapidité des instruments de calcul a, en un demi-siècle, été plusieurs fois multipliée par mille, cette conclusion n'est pas à prendre ironiquement.

Présentement, la physique théorique des hautes énergies progresse, principalement, en constituant des corpus de phénomènes rares parmi des ensembles immenses de cas ordinaires. La seule observation d'un de ces cas ordinaires demande des appareils de détection où jouent simultanément des millions de petits détecteurs élémentaires...

Finalement toute la Physique est subordonnée au progrès de cette branche particulière, de physique du solide, à très basse énergie, qui produit les outils de détection et de calcul. [Les "puces", dont le nom est connu de tous, n'étant que la moindre chose en la matière...].

Quant aux problèmes de l'avenir dans des domaines mieux connus de tous que la physique théorique et l'analyse des fonctions zeta... mais interdits jusqu'à présent à tout traitement numérique satisfaisant: analyse des images ou même seulement des sons de la musique et de la parole; configurations météorologiques saisies globalement, dans toutes leurs dimensions, alors que, de par le seul fait des mouvements du terme puisque chaque jour, de par les variables astronomiques, ne peut que différer des autres jours qui le suivent ou le précèdent... voici ma première impression.

Les praticiens foncent dans les analyses, les transformations de Fourier d'images, etc., sans savoir ce qu'ils cherchent.

Je suis assez satisfait des idées que je me suis faites sur la parole (même si j'ai plutôt un programme de recherches que des résultats suffisants: voir "Analyse spectrale et analyse statistique de la voix humaine parlée", *Les Cahiers de l'Analyse des Données*, Vol. XIII, 1988, no. 1, pp. 99–130). Je dois avouer (ne le dites à personne, je vous en prie!) que l'analyse des données n'y est pour rien. Ce qu'il faut, ce à quoi je me targue d'avoir quelque peu réussi... c'est voir ce qui, dans les objets étudiés, en l'espèce des sons, est pertinent. Voilà ce dont on doit d'abord s'enquérir dans tout corpus de dimension et de complexité "astronomiques".

Je le répète: le statisticien doit être modeste... le travail de ma génération a été exaltant... une nouvelle analyse est à inventer, maintenant que l'on a, et parfois à bas prix, des moyens de calcul dont on ne rêvait même il y a trente ans... Même si les voies explorées, jusqu'ici, dans certains domaines, sur lesquels il ne serait pas charitable d'insister, offrent à notre ironie une facile matière...

Jean-Paul Benzécri

Foreword

In 1954-1955, when I was visiting in Princeton, there was no computer at the university, nor at the Institute for Advanced Studies. A student could well devote a year to building an analog calculator, aimed at solving technical problems of a particular sort. It was the same in the Physics Lab of the École Normale Supérieure.

I felt that computers had to be marvellous devices. But, since they were available to so few, it was best to have recourse to extreme mathematical simplifications, so as to constrain the calculations within the limits of what was possible.

Thus, from 1959 to 1960 in the Operations Research Group of the National Marine, I had no scruples in reducing all multidimensional data to a normal distribution, as I collaborated from time to time with a colleague who looked after graphical simulations.

However, later, on the project of Automatic Translation of Natural Languages, linguistics, logic and mathematics came together in collaboration with the computer as tool, and it became clear that, on the path opened up by Louis Guttman † and Chikio Hayashi †, the principle of distributional equivalence proposed by the linguist Zelig Harris † ought to have been able to direct the analysis of statistical data.

Hence, giving geometrical form to this analysis, one would end up with the search for principal axes of inertia of a cloud of points endowed with mass. This is a classical problem in a dimensionality of 3, but is to be handled now in an arbitrary dimensionality of n. This necessarily requires diagonalizations of square $n \times n$ matrices, which was a calculation that was not feasible on a machine once n went beyond 3 or 4.

Around 1963, diagonalizing a 7×7 matrix was, for a lab with modest equipment, a significant task. Later, hierarchical clustering required even more heavy calculation compared to diagonalization. With machine capability increasing with algorithmic effectiveness, our career as statistician developed... using techniques, whose progress defied all dreams, servicing ambitions which grew without end!

Around 2000, on a PC that can be bought in a local store, one can in a few minutes cluster many thousands of observations. More exactly, a few minutes are needed for clustering algorithms and factor analysis. However the perception of the data, their formatting and preprocessing, the study of the results, – all of these take not just hours but months of work.

Strictly speaking, there is no longer any problem of computation. However the problem itself of data analysis persists. This is all the more vast in that computation no longer limits research. There is no excuse to stop with the collection of data and meditation. Relative to 1960, the difficulty ratio between intellectual plan and calculation has been inverted.

The principles though, that seem to us to hold now, are by no means accepted by all.

As far as the philosophy of numbers is concerned, the distinction between qualitative and quantitative seems to us to be still not always understood. In brief, it should not be held that:

- continuous numerical value $\approx$ quantitative value,

- value with a finite set of modalities $\approx$ qualitative value,

because at the level of the statistical item (e.g., patient's dossier) a numerical value (age, or even artery pressure or glycemy) is not generally to be taken with its given precision, but according to its significance. From this point of view, there is no difference between age and profession.

In particular, to compare an observation to another, one must consider not two sets of primary data (e.g., two sets of 100 real values, a point of $\mathbb{R}^{100}$, and another point of $\mathbb{R}^{100}$) between which global similarities are not in evidence, but the synthesis of these sets, ending up with a few gradations, or discontinuities, hence ultimately with diagnostics.

As far as computation is concerned, the factor analysis (for which, as we said, the basic complexity is matrix diagonalization) and hierarchical clustering algorithms, applied to data coded following the principle of distributional equivalence (i.e., profile) are far more effective for the interplay of neighbor relations than the stochastic approximation algorithms that are often carried out nowadays based on the schema of neural networks.

Such is the world as seen by a geometrician-statistician after 40 years of work. Is it allowed to assimilate the world to what one has observed and lived? This is certainly the usual case, but nonetheless reprehensible as a refusal to dream – at least (for want of doing even better than that) – of the future!

So, firstly: the issue of recognizing the order in what is contained in the elements perceived by the senses (or the tools that arm the senses) is perhaps simply philosophy in all its nobility. One says, in Latin, that all knowledge starts with the sensible in objects of nature: "Omnis cognitio initium habet a naturalibus... vel: a sensibilibus." Beyond such knowledge there is only mystery; and even revelation, given by God, is mediated by the example of what one knows from the natural play of reason.

Here the statistician, the geometrician, the sociologist, all must be modest! In searching for what one always had to search for, each generation can only play its part. The fundamental issue remains.

Secondly: as is clear from mathematics, the new computation for which the speed surpasses that of the computation of our generation by a ratio that is equally unimaginable today as it was a half century ago in the ratio of computational complexity that we were able to address relative to previous computation... one sees, I stress, that one should not say disparagingly that thought becomes lazy when the tool becomes more powerful.

On the one hand, to solve problems of integral calculus, "special functions" were invented. This made the way for the analysis of functions of complex variables (or at least furthered this analysis). Similarly for the integration of partial differential equations, precisely the theory of functional spaces is needed. Today, all calculations seem to be reduced to the basic play of the simplest methods, on networks of arbitrarily dense points.

In summary, the practical problem provokes (or at least spurs on) the development of theoretical ideas. The perfectioning of the tools makes theoretical speculation lazy.

However the reverse flow is also present. We find coincidences, and we give to these coincidences the shape of laws, put into circulation, before the development of appropriate theoretical ideas allows these laws to be proven.

The theory of analytical functions owes much to the desire to prove the great theorem of Fermat: $x^n + y^n = z^n$ does not have a solution for integers $n > 2$. Now Fermat was only able to conjecture after having first computed – noted – tried to disprove his statement or to prove it, and this was done with the means of computation of his time.

Here is an example: the abstract of a high energy theoretical physics article: hep-th/9811173, reproduced below. The abstract will interest even those who are no more aware than I am of the theory of generalized zeta functions and of the analysis of Feynman diagrams (from quantum field theory) carried out by counting vertices.

Following formidable computations, the authors end up with the result that fractions for which the numerator a can be of the order of a million, does not have a denominator b greater than 9. At the end there is this conclusion: *Our results are secure, numerically, yet appear very hard to prove by analysis.*

"Secure" is to be understood in statistical terms. If we use random sampling, there are, as it were, no sets of fractions of this sort giving rise to the arithmetic properties that were found.

We would be maybe even more surprised to find, in this same domain of high energy physics, conclusions like the following: "The result of the present work is not in doubt... However to decide on another issue ... we would need computations that are one thousand times more complex; and such calculations are currently impossible."

However since, we repeat, the rapidity of instruments of computation have been, in a half century, multiplied by one thousand many times over, this conclusion is not to be taken in an ironic way.

Currently, high energy theoretical physics progresses, mainly, by constituting corpora of rare phenomena among immense sets of ordinary cases. The simple observation of one of these ordinary cases requires detection apparatus based on millions of small elementary detectors.

Finally, all of physics is subordinate to progress in the particular branch of solid state, very low energy, physics, which produces detection tools and computation tools. The computer chips, known to everyone, are not the least significant manifestation of this.

Coming now to problems of the future, in domains that are more well known to all than theoretical physics, and the analysis of zeta functions... but inaccessible up to now to any satisfactory numerical treatment: the analysis of images, or even music and speech; or meteorological configurations taken globally, in all their dimensions. With the movement just of the moon and the sun there is no statistical population there, in the usual sense of the term, since each day has to differ from the previous one or the following one on account of different astronomical variables... That is my first impression.

Practitioners walk straight into the analyses, and transformations like Fourier analysis of images, without knowing what they are looking for. I do not have a personal experience of image analysis in dimensions of 2 or greater. It seemed that I was not equipped to look into that. But I did pass some months on a MacPlus, using a little "soundcap" software package, working on speech. This resulted in the article "Analyse spectrale et analyse statistique de la voix humaine parlée" ("Spectral and statistical analysis of the spoken human voice") in *Les Cahiers de l'Analyse des Données*, vol. XIII, 1988, pp. 99–130.

I am sufficiently satisfied with the ideas that I developed on speech (even if I have more a program of research than sufficient results). I must admit (keep it quiet, please!) that data analysis does have its uses. What is needed, and what I am proud about having somewhat succeeded in doing, is to see what is relevant in the objects studied, in the species of sounds. That is what one should first look into, in any corpus of "astronomical" dimension and complexity.

I repeat: the statistician has to be modest. The work of my generation has been exalting. A new analysis is there to be invented, now that one has inexpensive means of computation that could not be dreamed of just thirty years ago. To be charitable, we say this even if the paths explored up to now, in certain domains, allowed us easily to be ironic about them.

Jean-Paul Benzécri

hep-th/9811173
Determinations of rational Dedekind-zeta invariants of hyperbolic manifolds and Feynman knots and links
Authors: *J.M. Borwein, D.J. Broadhurst*
Comments: *53 pages, LaTeX*
Report-no: *OUT-4102-76; CECM-98-120*
Subj-class: *High Energy Physics - Theory; Geometric Topology; Number Theory; Classical Analysis and ODEs*
We identify 998 closed hyperbolic 3-manifolds whose volumes are rationally related to Dedekind zeta values, with coprime integers a and b giving $a/b\mathrm{vol}(M) = (-D)^{3/2}/(2\pi)^{2n-4}$ $(\zeta_K(2))/(2\zeta(2))$ for a manifold M whose invariant trace field K has a single complex place,

discriminant D, degree n, and Dedekind zeta value $\zeta_K(2)$. The largest numerator of the 998 invariants of Hodgson-Weeks manifolds is, astoundingly, $a = 2^4 \times 23 \times 37 \times 691 = 9,408,656$; the largest denominator is merely $b=9$. We also study the rational invariant a/b for single-complex-place cusped manifolds, complementary to knots and links, both within and beyond the Hildebrand-Weeks census. Within the censi, we identify 152 distinct Dedekind zetas rationally related to volumes. Moreover, 91 census manifolds have volumes reducible to pairs of these zeta values. Motivated by studies of Feynman diagrams, we find a 10-component 24-crossing link in the case $n=2$ and $D=-20$. It is one of 5 alternating platonic links, the other 4 being quartic. For 8 of 10 quadratic fields distinguished by rational relations between Dedekind zeta values and volumes of Feynman orthoschemes, we find corresponding links. Feynman links with $D=-39$ and $D=-84$ are missing; we expect them to be as beautiful as the 8 drawn here. Dedekind-zeta invariants are obtained for knots from Feynman diagrams with up to 11 loops. We identify a sextic 18-crossing positive Feynman knot whose rational invariant, $a/b=26$, is 390 times that of the cubic 16-crossing non-alternating knot with maximal D_9 symmetry. Our results are secure, numerically, yet appear very hard to prove by analysis.

About the Author

Fionn Murtagh studied with Jean-Paul Benzécri, completing his doctorate in January 1981. He holds the post of Professor of Computer Science at the University of London. He is also Adjunct Professor at Strasbourg Astronomical Observatory, Strasbourg. He has published extensively on data analysis and signal processing. He is a Member of the Royal Irish Academy, and a Fellow of the British Computer Society.

Contents

1

Introduction

In this first chapter we indicate the particular orientation of this book. We then take a very broad historical perspective on data analysis. An illustrative example takes as its point of departure the closely related core algorithms used in principal components analysis and in correspondence analysis. From contrasting the results we show how and where the two approaches respond to different exigencies. Finally we present R software – which is handy as pseudo-code description as well as to be run – that was used in the illustrative example.

1.1 Data Analysis

In this work we aim to present in a new way some of the most central and most important aspects of Jean-Paul Benzécri's data analysis framework or system. Our contributions include the following:

- *Describing the many high points of an integrated approach to analyzing and understanding data that was developed and finessed over four decades by Benzécri's "French School" of data analysis.*

- *From the point of view of the underlying mathematics, expressing data analysis as a tale of two metrics, Euclidean and χ^2 (chi squared).* Our familiar Euclidean framework can be used for visualization and for summarization. But the χ^2 metric is an ideal one for handling many types of input data.

- *From the practical point of view, focusing on the crucial area of input data coding.* When one considers that 0 and 1 codes are at the basis of computer technology, or that a four letter code is at the basis of the genes that make up the human or animal, then it is clear that data coding goes right to the heart of all science and technology. In fact data coding is part and parcel of our understanding of observed reality.

- *Providing software code in widely used languages, and discussing its use throughout the book.* This we do for our code in R, a public domain statistical and graphical data analysis system that is available for all platforms. See www.r-project.org, where an easy to install binary version of R can be obtained for all Windows (Intel) systems, Linux, Solaris (Sun) and Mac OS X. In addition, we describe a system that runs as a Java application, with support for correspondence analysis as such, hierarchical clustering, and aids to mutual interpretation, together with basic support for input and graphical and tabular output.

This mathematical and algorithmic framework is centered on correspondence analysis, and more or less equally on clustering and classification. We will cover all salient aspects of the principles behind these algorithms.

We will detail a wide range of applications and case studies, some of which are reworked from past work of ours or of others, and some of which are new and have not been published before.

The applications have been carried out using our own software, which constitutes a very important part of this book. This software has been written in two separate ways, that are parallel in a sense: in Java and in R. Java has become an important programming language and is widely used in teaching and in cross-platform (desktop and handheld computing platforms, Windows or Unix operating systems where the latter includes Macintosh), programming development. R is an open source version of S-Plus, which took further the S statistical and graphical programming language. These programs were written by the author. The correspondence analysis and "aids to interpretation" programs try to follow fairly closely the programs in use in J.-P. Benzécri's laboratory through the 1970s, 1980s and 1990s. The hierarchical clustering programs distill the best of the reciprocal nearest neighbors algorithm that was published around 1980 in J.-P. Benzécri's journal, *Les Cahiers de l'Analyse des Données* ("Journal of Data Analysis") and have not been bettered since then.

Much of the development work in this framework, ranging from Einstein tensor notation through to the myriad application studies published in the journal *Les Cahiers de l'Analyse des Données*, were advanced by the time of my arrival in Paris in late 1978 to start course work and embark on a doctoral thesis with J.-P. Benzécri. At that time I obtained the two central Benzécri tomes on *L'Analyse des Données* Volume 1 – *La Taxinomie*, and Volume 2 – *L'Analyse des Correspondances*, 1976 second editions, published by Dunod. Ever since, I have valued these most elegant volumes, packed full of theory and examples, and even Fortran code. They are wonderful books. With full page line drawings, they are placed, respectively, under the aegis of Carl Linnaeus (or Carl von Linné, who lived from 1707 to 1778 in Sweden, and is known as the Father of Taxonomy), and of Christian Huyghens (who lived from 1629 to 1695 in The Netherlands, and who made fundamental contributions to mathematics and mechanics).

Now the sleeve of the second Volume is simply noted *Correspondances*. It is very fitting therefore to hotlink to the first verse of the poem of this name by Charles Baudelaire, great lyric poet of nineteenth century France, in his epic work, *Les Fleurs du Mal*. One observes the temple of nature is a way that is certainly not overly clear, as any data analyst or observational scientist appreciates. Understanding is not possible without recourse to "forests" of mathematical and algorithmic symbols and tools. Hacking a way through the forests of symbols is therefore a necessary task for the data analyst, in order to get close to that which is studied.

Correspondances

> La nature est un temple où de vivants piliers
> Laissent parfois sortir de confuses paroles;
> L'homme y passe à travers des forêts de symboles
> Qui l'observent avec des regards familiers.

("Nature is a temple whose live pillars sometimes emit confused expression; man gets to this temple through forests of symbols, who observe him with familiar looks.")

1.2   Notes on the History of Data Analysis

A little book published in 1982, *Histoire et Préhistoire de l'Analyse des Données* [19] – we will refer to it as the *Histoire* – offers nice insights into multivariate data analysis or multidimensional statistics. But it does a lot more too. It gives a short but fascinating historical sweep of all of observational data analysis. It was written in the spring of 1975, circularized internally, published chapter-wise in *Les Cahiers*, before taking book form.

The first chapter takes in early theory of probability and doctrine of chance, Bayes, theory of error, the normal or Gaussian distribution which was anticipated by Laplace by some 40 years over Gauss, linkages with kinetic theory of gases, paradoxes in probability and their resolution.

It begins with a theme that echoes widely in Benzécri's writings: namely that the advent of computers have overturned statistics as understood up until then, and that the suppositions and premises of statistics have to be rethought in the light of computers. From probability theory, data analysis inherits inspiration but not methods: statistics is not, and cannot be, probability alone. Probability is concerned with infinite sets, while data analysis only touches on infinite sets in the far more finite world expressed by such a typical problem as discovering the system of relationships between rows and columns of a rectangular data table.

1.2.1   Biometry

The next major trend in statistical work is linked to Quetelet who sought to ground the work of Charles Darwin in statistical demonstrations. Quetelet worked on anthromorphology, and later Galton and Pearson worked on biometry. Adolphe Quetelet was a Belgian astronomer, whose work extended into meteorology and biometry. The work of Gauss and Laplace had been centered on the principle of least squares. In the biometry work this was extended to take in correlation (by Auguste Bravais, French physicist, astronomer and polymath) and regression.

The principal axes of a cloud of multidimensional points were known to Bravais. Karl Pearson and R. Weldon (who died at the age of 46) at University College London developed biometrics in the twentieth century. With an article in 1901 in the Philosophical Magazine entitled "On lines and planes of closest fit to systems of points in space," Pearson is the clear originator of principal components analysis and related factor analysis methods. The term "factor" to denote a new underlying variable comes from genetics. Pearson was to be followed and somewhat outshone by Fisher who made statistical methodology more coherent but at the cost of being less flexible in regard to observational data. For Benzécri the patronage of Pearson is to be preferred!

1.2.2   Era Piscatoria

The wave of biometry was followed by the wave of agronomy – with genetics underlying both – whose chief exponent was Sir Ronald Fisher. Yule said, when Karl Pearson died in 1936, in a word-play on Fisher's name, "I feel as though the Karlovingian era has come to an end, and the Piscatorial era which succeeds it is one in which I can play no part."

Fisher, for Benzécri, was bettered by no one in his sense of ease in multidimensional spaces, but was overly devoted to exact mathematical distributions, and most of all the normal or Gaussian distribution.

Using Fisher's admirable geometrical perspective a short explanation is given of Student's (Gosset's) t-distribution, oriented towards using statistical quantities defined for large numbers of observations when relatively few observations were to hand.

Non-parametric approaches based on ranks are touched on: Fisher was not too keen on these, and neither is Benzécri insofar as a cloud of points is not easily handled in terms of rank data. But situations of importance for rank order data include psychometrics (more on which below) and general validation experiments that can be counterposed and related to a data analysis.

Maximum likelihood estimation is discussed next, touching on use of Bayesian priors, in a geometric framework.

This is followed by analysis of variance: a data decomposition to suggest causes and effects (particularly relevant in agronomy), which differs from the data decomposition used in correspondence analysis (which is more general:

the "only possibility, we believe, in most disciplines"). Examples of experimental design are discussed.

Discussion of Fisher's work in the data analysis context would not be complete without discussion of discriminant analysis. In developing discriminant analysis – in later chapters we will use the iris flower data set that has become a classic data set and that was used by him in his 1936 article that established Fisher's linear discriminant analysis [3, 42] – Fisher in fact developed the basic equations of correspondence analysis but without of course a desire to do other than address the discrimination problem. Discriminant analysis, or supervised classification, took off in a major way with the availability of computing infrastructure. The availability of such methods in turn motivated a great deal of work in pattern recognition and machine learning.

Deduction means to derive general principles from particular consequences; induction means to raise from knowledge of particular facts up to general concepts and the links between these concepts. In defense of real world applications, Fisher set himself against deduction. So in a way does Benzécri, in saying that "statistics is not probability." Now, though, Benzécri argues for a more nuanced view, and lack of any strict opposition between deductive and inductive (e.g., inference and decision-oriented) views.

Again it is noted that computer-based analysis leads to a change of perspective with options now available that were not heretofore. Fisher's brilliant approaches implicitly assume that variables are well known, and that relations between variables are strong. In many other fields, a more exploratory and less precise observational reality awaits the analyst.

1.2.3 Psychometrics

Earlier biometry work evolved into psychometrics (psychological measurement). Indeed correspondence analysis did also, in that it arose from early work on language data analysis, analysis of questionnaires, and textual analysis. With the computer, convergence takes place between on the one hand psychometrics, and on the other hand "the taxonomic constructions of the ecologists, and the ambitious dreams of pattern recognition."

Psychometrics made multidimensional or multivariate data analysis what it has now become, namely, "search by induction of the hidden dimensions that are defined by combinations of primary measures." Psychometrics is a response to the problem of exploring areas where immediate physical measurement is not possible, e.g., intelligence, memory, imagination, patience. Hence a statistical construction is used in such cases ("even if numbers can never quantify the soul!").

Psychophysics, as also many other analysis frameworks such as the method of least squares, was developed in no small way by astronomers: the desire to penetrate the skies led too to study of the scope and limits of human perception. Later, around the start of the 20th century, came interest in human intelligence, and an underlying measure of intelligence, the intelligence quotient (IQ). From the 1930s, Louis Leon Thurstone developed factor analysis.

Having now touched on factor analysis, we will take a brief detour from Benzécri's *Histoire* to comment on various connotations of terms used in the area of data analysis. Firstly, the term "analyse des données" in French, because of Benzécri's work, came from the 1980s onwards (maybe earlier) to mean (i) principled analysis procedures anchored around correspondence analysis, with (ii) ancillary analysis procedures, in the first instance hierarchical clustering, and often enough discriminant analysis, maybe with regression analysis, and certainly graphics and display. On the other hand the term "multidimensional" or "multivariate data analysis" in English can mean somewhat different things: (i) purely exploratory and graphical analysis, often associated with the name of Tukey (the great statistician whose career was spent at Princeton, 1915–2000, who gave us the word "bit" meaning "binary digit"; and, with J.W. Cooley, the Fast Fourier transform algorithm); or (ii) analysis anchored around the multivariate Gaussian or normal distribution; or (iii) interactive, statistical graphics. So much for the term "data analysis."

Next let us look at the term "analyse factorielle" in French. In the Benzécri tradition this means correspondence analysis or principal components analysis. But in English, "factor analysis," associated with Thurstone and others, means an analysis method somewhat like principal components analysis but involving "commonality" values on the diagonal of the variance/covariance or correlation matrix; rotation of factors that are, when found, orthogonal; and so on.

Maurice Kendall (statistician, 1907–1983) described principal components analysis as going from data to model, while factor analysis goes from model to data. Benzécri does not agree, and nor also did Thurstone when he admitted in his 1947 book on *Multiple Factor Analysis* that "Factor analysis has not been generally accepted by mathematical statisticians..." Indeed if one accepts that Thurstonian factor analysis is not really at all assessing model fit to data, then it is quite reasonable to conclude that model-oriented statisticians are ill at ease with factor analysis.

On this note we will come to the final terms that we will briefly discuss: "data analysis" versus "statistics." For some the two are nearly equal; and for other scientists they represent two different views of the world.

We now return to our discussion of some of the high points of Benzécri's *Histoire*.

"Particularly neat and simple mathematical properties" was Torgerson's justification for use of Euclidean distance in data analysis. Torgerson [81] used a continuum for judgements and perception. "Discriminal process" was a point on a continuum. Guttman continued this work and applied it to the assessment of US army recruits in the Second World War. Guttman's models, and Lazarsfeld's latent structures, came close to the type of formalism and notation that is part and parcel of correspondence analysis. Perspectives on data analysis were synthesized further by Coombs in his *A Theory of Data* in 1964 [34]. The work of Coombs is much appreciated by Benzécri, even if preference data are considered as playing too prominent a role relative to other types of data. Coombs is taken as representing a "period of transition" before the advent of accessible computer infrastructure.

With computers came pattern recognition and, at the start, neural networks. A conference held in Honolulu in 1964 on *Methodologies of Pattern Recognition* cited Rosenblatt's perceptron work many times (albeit his work was cited but not the perceptron as such). Early neural network work was simply what became known later as discriminant analysis. The problem of discriminant analysis, however, is insoluble if the characterization of observations and their measurements are not appropriate. This leads ineluctably to the importance of the data coding issue for any type of data analysis.

1.2.4   Analysis of Proximities

Around the mid-1960s Benzécri began a correspondence with R.N. Shepard which resulted in a visit to Bell Labs. Shepard ("a statistician only in order to serve psychology, and a psychologist out of love for philosophy") and D. Carroll (who "joyfully used all his ingenuity – which was large indeed – to move data around in the computer like one would move pearls in a kaleidoscope") had developed proximity analysis. These were algorithms to handle the analysis of experimental associations between stimulus and response. Just like in correspondence analysis, a central aim is to map the data into a low-dimensional representation that can be visualized easily. Multidimensional scaling does this using an iterative approximation approach.

A back-of-envelope approach to finding a low-dimensional fit to the data was developed by Benzécri and is described in the volume, *La Taximonie* [15]. If the points, in the original space of large dimension, m, are distributed following a normal or Gaussian distribution then their squares follow a χ^2 distribution with m degrees of freedom. This is well-known from basic statistical theory. As is usual in proximity analysis, ranks of proximities – i.e., inversely associated with distances or (less rigorous) dissimilarities – are used. Say that we have 9 points, which leads to $9 \cdot (9-1)/2 = 36$ distances to be estimated. We target a two-dimensional representation of these 9 points, so the number of degrees of freedom in the χ^2 distribution is 2. Now the cumulative χ^2 distribution with 2 degrees of freedom is a curve with, on the horizontal axis increasing values of distance squared, and on the vertical axis the χ^2 value from 0 to 1. What we do is use the rank, transformed to be between 0 and 1, on the vertical axis, throw a line across to the cumulative χ^2 curve, and then throw a line down to the horizontal axis. This allows us to read off a value for the distance squared. To transform the rank to be between 0 and 1, we used for our chosen example $(37 - r)/37$ where r is the rank of our dissimilarity (viz., between 1 and 36). The distance squared that we find has a probability of $r/37$ of having a greater than or equal χ^2 distribution value. With our distance squared, we take our points in sequence and with a compass place them on a two-dimensional plot.

1.2.5   Genesis of Correspondence Analysis

The term "correspondence analysis" was first proposed in the fall of 1962. The first presentation under this title was made by J.-P. Benzécri at the Collège de France in a course in the winter of 1963.

By the late 1970s what correspondence analysis had become was not limited to the extraction of factors from any table of positive values. It also catered for data preparation; rules such as coding using complete disjunctive form; tools for critiquing the validity of results principally through calculations of contribution; provision of effective procedures for discrimination and regression; and harmonious linkage with cluster analysis. Thus a unified approach was developed, for which the formalism remained quite simple, but for which deep integration of ideas was achieved with diverse problems. Many of the latter originally appeared from different sources, and some went back in time by many decades.

Two explanations are proposed for the success of correspondence analysis. Firstly, the principle of distributional equivalence allows a table of positive values to be given a mathematical structure that compensates, as far as possible, for arbitrariness in the choice of weighting and subdivision of categories. Secondly, a great number of data analysts, working in very different application fields, found available a unified processing framework, and a single software package. Correspondence analysis was considered as a standard, unifying and integrated analysis framework – a platform.

Correspondence analysis was initially proposed as an inductive method for analyzing linguistic data. From a philosophy standpoint, correspondence analysis simultaneously processes large sets of facts, and contrasts them in order to discover global order; and therefore it has more to do with synthesis (etymologically, to synthesize means to put together) and induction. On the other hand, analysis and deduction (viz., to distinguish the elements of a whole; and to consider the properties of the possible combinations of these elements) have become the watchwords of data interpretation. It has become traditional now to speak of data analysis and correspondence analysis, and not "data synthesis" or "correspondence synthesis."

N. Chomsky, in the little volume, *Syntactic Structures* [33], held that there could not be a systematic procedure for determining the grammar of a language, or more generally linguistic structures, based on a set of data such as that of a text repository or corpus. Thus, for Chomsky, linguistics cannot be inductive (i.e., linguistics cannot construct itself using a method, explicitly formulated, from the facts to the laws that govern these facts); instead linguistics has to be deductive (in the sense of starting from axioms, and then deriving models of real languages).

Benzécri did not like this approach. He found it idealist, in that it tends to separate the actions of the mind from the facts that are the inspiration for the mind and the object of the mind. At that time there was not available an effective algorithm to take 10,000 pages of text from a language to a syntax, with the additional purpose of yielding semantics. But now, with the advances in our computing infrastructure, statistics offers the linguist an effective inductive method for usefully processing data tables that one can immediately collect, with – on the horizon – the ambitious layering of successive research that will not leave anything in the shade – from form, meaning or style.

This then is how data analysis is feasible and practical in a world fuelled by computing capability: "We call the distribution of a word the set of its possible environments." In the background we have a consideration that Laplace noted: a well-constructed language automatically leads to the truth, since faults in reasoning are shown up as faults in syntax.

In response to the structuralists who hold that objects do not exist and instead only relations between the objects really exist, Benzécri in replying echoes Aristotle's *On the Soul* that objects do exist, but they are only revealed to us through their relations.

From 1950 onwards, statistical tests became very popular, to verify or to protect the acceptability of a hypothesis (or of a model) proposed a priori. On the other hand correspondence analysis refers from the outset to the hypothesis of independence of observations (usually rows) I and attributes (usually columns) J but aims only at exploring the extent to which this is not verified: hence the spatial representation of uneven affinities between the two sets.

Correspondence analysis looks for typical models that are achieved a posteriori and not a priori. This is following the application of mutual processing of all data tables, without restrictive hypotheses. Thus the aim is the inductive conjugating of models. The initial project motivated by the Chomskyian thesis was carried out: viz., to give a formalized method to induction. This was truly an old ambition, spoken about by Bacon around 1600. In homage to Bacon's Novum Organum (*The New Organon or True Directions Concerning the Interpretation of Nature*, [11]), Benzécri proposed that we should term data analysis the Novius ("Newer") Organum.

1.3 Correspondence Analysis or Principal Components Analysis

1.3.1 Similarities of These Two Algorithms

Principal components analysis is a data analysis algorithm that is similar to correspondence analysis. At least the core visualization objective is shared by both. They both start with a data table or array as input. They both construct a low-dimensional Euclidean output, that can be visualized as a type of map. The mathematics behind both is also very similar: the factors in correspondence analysis, and the principal components in principal components analysis, are defined from the eigenvectors of a square, symmetric matrix. This square matrix is often a correlation matrix in principal components analysis. In correspondence analysis, the square matrix is also a cross-products matrix, and we will leave further details of this to later chapters.

1.3.2   Introduction to Principal Components Analysis

Principal components analysis comes in three "flavors," and – to make life simple for the user – we would suggest that the third one is the most important in practice. The same basic algorithm applies to all three "flavors" but they differ in data preprocessing, prior to the determining of the principal components. Say that the input data array is denoted X, which contains values which cross n observations with m variables. Each observation can be considered as a point in an m-dimensional space. Each variable, analogously, can be considered as a point in an n-dimensional space.

The variables used may be very different in value: some may "shout louder" than others. To enforce a measure of homogeneity on our data, we may center the observations, and we may "reduce" them to unit variance.

To center the observations, say that the ith observation has values that are represented thus: $x_{i1}, x_{i2}, x_{i3}, \ldots, x_{im}$. Let the mean observation be given by: $\bar{x}_1, \bar{x}_2, \bar{x}_3, \ldots, \bar{x}_m$. Centering each observation gives: $x_{i1} - \bar{x}_1, x_{i2} - \bar{x}_2, x_{i3} - \bar{x}_3, \ldots, x_{im} - \bar{x}_m$. We have $\bar{x}_1$ defined as $\bar{x}_1 = 1/n \sum_i x_{i1}$. This means that we have completely relocated the origin or zero point in our space. There are major implications for principal components analysis. We will be seeking a new system of axes, to better fit our data. The new system of axes will have a new, and better, origin.

Reduction to unit variance involves rescaling the original variables in the following way. Let us assume that centering has already taken place. Then we transform: $x_{i1} - \bar{x}_1, x_{i2} - \bar{x}_2, x_{i3} - \bar{x}_3, \ldots, x_{im} - \bar{x}_m$ to the following: $(x_{i1} - \bar{x}_1)/\sigma_1, (x_{i2} - \bar{x}_2)/\sigma_2, (x_{i3} - \bar{x}_3)/\sigma_3, \ldots, (x_{im} - \bar{x}_m)/\sigma_m$. The variance is σ^2, so we are using the standard deviation here. It is defined as: $\sigma_1^2 = 1/n \sum_i (x_{i1} - \bar{x}_1)^2$. The effect of reduction to unit variance is to enforce similar intervals on all variables.

There are many other ways that we can enforce homogeneity on the input data. Motivation for centering to zero mean, and standardizing to unit variance, is linked to seeing the data as an approximately Gaussian-shaped cloud of points. With this viewpoint there is no asymmetry to be considered, nor multimodality, nor outlier values.

Further important motivation for centering to zero mean, and standardizing to unit variance, is linked to seeing the data values as realizations of a continuous distribution of values.

Both of these motivations are different in correspondence analysis. Firstly, the enforcing of homogeneity is carried out differently, as will be seen in later chapters. Secondly, the data that is preferably handled by correspondence analysis includes data tables containing frequencies, or scores, or presence/absence values, and other kinds of qualitative or categorical values.

The "flavors" of principal components analysis can now be listed. Let X be the initial data. Principal components analysis takes place on $X^t X$ and in the dual space (e.g., space of variables, if we had considered the space of observations to begin with) $X X^t$ where ^{t} denotes transpose.

1. If X is neither centered to zero mean, nor reduced to unit variance, then $X X^t$ is a so-called sums of squares and cross-products matrix.

2. If X is centered to zero mean, then $X X^t$ is the variance-covariance matrix.

3. If X is both centered to zero mean, and reduced to unit variance, then $X X^t$ is the correlation matrix.

Different analyses result from these "flavors." Generally the last one, the analysis of correlations, can be recommended.

1.3.3 An Illustrative Example

To illustrate the types of visualizations which will be studied in greater depth elsewhere in this book, we will take a small data set used in [31]. The authors collected data on 14 different galactic globular clusters – dense masses of stars, all of which had been collected in earlier CCD (charge coupled device, – digital detector) photometry studies. They studied a number of salient associations between variables. We will do this based on global simultaneous views of the data, and briefly note some aspects of these analyses.

The variables used in Table 1.1 are, respectively: relaxation times, galactocentric distance, distance to galactic disk, logarithm of total mass, concentration parameter, metallicity, and local (observed) and global (after mass-segregation effect corrections) power law index of fit to slope of the mass function. Distance is measured in Kpc, kilo parsecs; M. denotes solar mass.

A set of all pairwise plots is shown in Figure 1.1. Each variable is plotted against another. The plot panels are symmetric about the principal diagonal, so that each plot appears twice here – once as a transposed version. We could label the points. Not having done so means that correlations and clusters are essentially what we will instead look for. Consider the plots of variables x and x_0: they are very highly correlated, such that one or other of these variables is in fact redundant. Consider x and Z_g: the relationship is less spectacular, but nonetheless very correlated. This fact is discussed in [31]. Outliers, to be interpreted as anomalous observations or perhaps as separate classes, are to be seen in quite a few of the plot panels.

TABLE 1.1

Data: 14 globular clusters, 8 variables.

Object	t_rlx years	Rgc Kpc	Zg Kpc	log(M/ M.)	c	[Fe/H]	x	x0
M15	1.03e+8	10.4	4.5	5.95	2.54	-2.15	2.5	1.4
M68	2.59e+8	10.1	5.6	5.1	1.6	-2.09	2.0	1.0
M13	2.91e+8	8.9	4.6	5.82	1.35	-1.65	1.5	0.7
M3	3.22e+8	12.6	10.2	5.94	1.85	-1.66	1.5	0.8
M5	2.21e+8	6.6	5.5	5.91	1.4	-1.4	1.5	0.7
M4	1.12e+8	6.8	0.6	5.15	1.7	-1.28	-0.5	-0.7
47 Tuc	1.02e+8	8.1	3.2	6.06	2.03	-0.71	0.2	-0.1
M30	1.18e+7	7.2	5.3	5.18	2.5	-2.19	1.0	0.7
NGC 6397	1.59e+7	6.9	0.5	4.77	1.63	-2.2	0.0	-0.2
M92	7.79e+7	9.8	4.4	5.62	1.7	-2.24	0.5	0.5
M12	3.26e+8	5.0	2.3	5.39	1.7	-1.61	-0.4	-0.4
NGC 6752	8.86e+7	5.9	1.8	5.33	1.59	-1.54	0.9	0.5
M10	1.50e+8	5.3	1.8	5.39	1.6	-1.6	0.5	0.4
M71	8.14e+7	7.4	0.3	4.98	1.5	-0.58	-0.4	-0.4

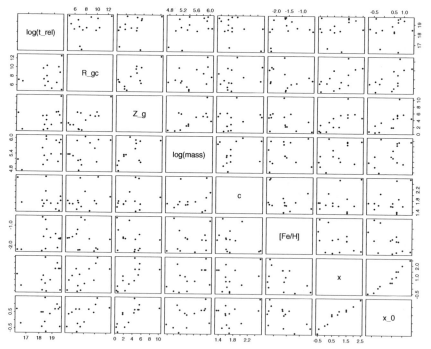

FIGURE 1.1

All pairwise plots between 8 variables.

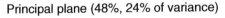

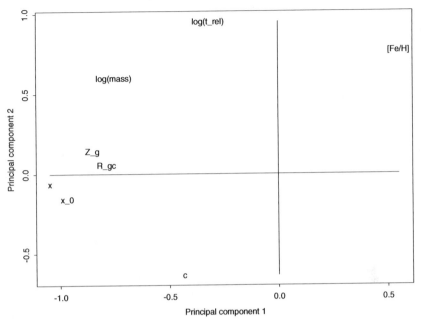

FIGURE 1.2
Principal component analysis (principal plane) of the 8 variables.

1.3.4 Principal Components Analysis of Globular Clusters

Figure 1.2 shows a principal components analysis of the variables. This is an optimal planar representation of the data (subject, of course, to what we mean by optimality: in Chapter 2 we will define this). The variable [Fe/H] is quite different from the others, in that it is quite separate in the plot from other variables. This variable itself is a ratio of iron to hydrogen in the stellar atmosphere and is a result of chemical evolution. It expresses metallicity and was discussed in [31] and used as a basis for subdividing the globular clusters. We could also view the latter in the principal plane and find, in that way, what observations are most positively associated with the variable [Fe/H].

Another way to visualize the observations, if clustering is really what we are interested in, is to directly cluster them. A hierarchical clustering provides lots of classification-related information. Figure 1.3 shows such a classification tree, or dendrogram. Two large clusters are evident, comprising the 6 globular clusters to the left, and the 8 globular clusters to the right. Note how the branches could be reversed. However what belongs in any given branch will not change, subject to the particular clustering criterion being used. In Figure 1.3, a criterion seeking maximal cluster compactness (defined by within cluster

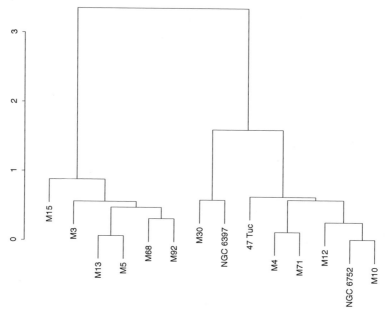

Hierarchical clustering (Ward's) of globular clusters

FIGURE 1.3

Hierarchical clustering of the 14 globular clusters.

variances) is employed.

These methods are relatively powerful. They allow us to answer questions related to internal associations and correlations in our data. They provide answers to degrees of redundancy and of "anomalicity." They provide visualizations to help us with communication of our conclusions to our clients or colleagues. They are tools (algorithmic, software) which are easy to use, and which let the data speak for themselves.

1.3.5 Correspondence Analysis of Globular Clusters

Logarithms have been applied to the data (in the fourth variable, as given, and in the first variable by us). From the pairwise plots we see that the distributions of variables are not Gaussian. We will apply the following fuzzy coding for use in correspondence analysis. Each variable value is replaced by a couple of values $(v, \bar{v})$ with the following properties. If the value is greater than the 66.6th percentile, then the couple $(1, 0)$ replaces the original value. If the value is less than the 33.3rd percentile, then the couple $(0, 1)$ replaces the original value. If the value is less than the 66.6th percentile, but greater than the 33.3rd percentile, then the couple $(v, \bar{v})$ will have values that are pro

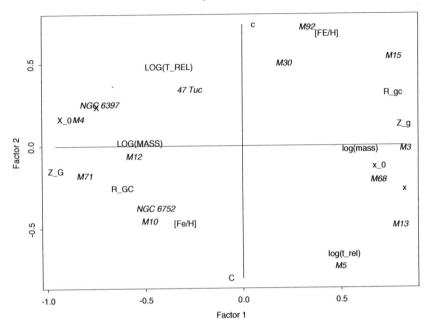

Principal factor plane

FIGURE 1.4

Correspondence analysis (principal factor plane) of the 16 fuzzily coded variables, and the 14 globular clusters.

rata, between 0 and 1. In all cases, $v + \bar{v} = 1$.

Figure 1.4 shows the correspondence analysis. There are now 16, rather than 8, variables. We have noted the "upper" partner among the couple of variables with the same label as before in principal components analysis; and the "lower" partner among the couple of variables with an upper case label. Many aspects of the analysis result displayed in this best-fitting planar representation are quite similar. For example, metallicity, [Fe/H], is quite separate and distinct in both analyses (Figures 1.2 and 1.4). This similarity in the display results is notwithstanding the fact that the input data to the two analyses are quite different.

The hierarchical clustering results can be more directly appreciated through Figures 1.3 and 1.5. The original data was used in the former case, and the fuzzily recoded data was used in the latter case. Generally, in correspondence analysis it is better to use the factor projections as input to the clustering. This is because the correspondence analysis includes data reweighting, and the output produced – in terms of factor projections – is Euclidean.

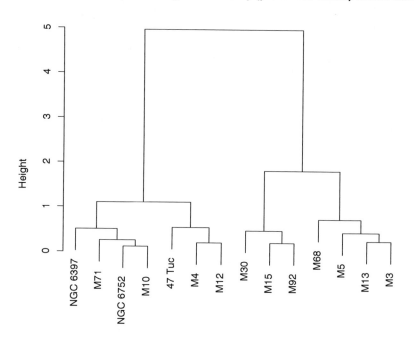

FIGURE 1.5

Hierarchical clustering of the 14 globular clusters, based on the 16 fuzzily coded variables.

1.4 R Software for Correspondence Analysis and Clustering

Software discussed here, and data sets, are available on the web at: www.correspondances.info .

A short discussion follows on the software used for the correspondence analysis. The language, or rather environment, R (www.r-project.org) is open source, with quickly installed binary versions available for Windows, Unix and Macintosh. R, we find, is convenient for data input and output, excellent for data display, and outstanding for the preparation of graphical materials for publications or presentations. Since it is a high level language it provides also a convenient description of operations.

1.4.1 Fuzzy or Piecewise Linear Coding

First the fuzzy coding which we apply to each variable in the given data array, `fuzzy.r`:

```
fuzzy <- function(x)
  {
    # Fuzzy or piecewise linear coding of a vector of values.

    if (!is.vector(x))
        stop("Only vectors handled currently as input.\n")
    plin.x <- cbind(x, x)

    lo     <- quantile(x, .33)
    hi     <- quantile(x, .67)

    plin.x[x >= hi, 1] <- 1
    plin.x[x >= hi, 2] <- 0

    plin.x[x <= lo, 1] <- 0
    plin.x[x <= lo, 2] <- 1

    plin.x[x > lo & x < hi, 1] <-
        (x[x > lo & x < hi] - lo) / (hi - lo)
    plin.x[x > lo & x < hi, 2] <-
        (hi - x[x > lo & x < hi]) / (hi - lo)

    plin.x

  }
```

1.4.2   Utility for Plotting Axes

A little utility, `plaxes.r`, for plotting axes is as follows:

```
plaxes <- function(a,b) {

        segments(   min(a),0,   max(a),0   )
        segments(   0,min(b),   0,max(b)   )

}
```

1.4.3   Correspondence Analysis Program

The correspondence analysis program, `ca.r`, is as follows.

```
ca <- function(xtab) {

# Correspondence analysis of principal table.  List returned
# with projections, correlations, and contributions of rows
# (observations), and columns (attributes).  Eigenvalues are
# output to display device.

tot <- sum(xtab)
fIJ <- xtab/tot
fI <- apply(fIJ, 1, sum)
fJ <- apply(fIJ, 2, sum)
fJsupI <- sweep(fIJ, 1, fI, FUN="/")
fIsupJ <- sweep(fIJ, 2, fJ, FUN="/")
s <- as.matrix(t(fJsupI)) %*% as.matrix(fIJ)
s1 <- sweep(s, 1, sqrt(fJ), FUN="/")
s2 <- sweep(s1, 2, sqrt(fJ), FUN="/")
# In following s2 is symmetric.  However due to precision S-Plus
# didn't find it to be symmetric.  And function eigen in S-Plus
# uses a different normalization for the non-symmetric case (in
# the case of some data)!  For safety, we enforce symmetry.
sres <- eigen(s2,symmetric=T)
sres$values[sres$values < 1.0e-8] <- 0.0
cat("Eigenvalues follow (trivial first eigenvalue removed).\n")
cat(sres$values[-1], "\n")
cat("Eigenvalue rate, in thousandths.\n")
tot <- sum(sres$values[-1])
cat(1000*sres$values[-1]/tot,"\n")
# Eigenvectors divided rowwise by sqrt(fJ):
evectors <- sweep(sres$vectors, 1, sqrt(fJ), FUN="/")

# PROJECTIONS ON FACTORS OF ROWS AND COLUMNS
```

```
rproj <- as.matrix(fJsupI) %*% evectors
temp  <- as.matrix(s2) %*% sres$vectors
# Following divides rowwise by sqrt(fJ) and columnwise by
# sqrt(eigenvalues):
# Note: first column of cproj is trivially 1-valued.
# NOTE: Vbs. x factors.
# Read projns. with factors 1,2,... from cols. 2,3,...
cproj <- sweep(sweep(temp,1,sqrt(fJ),FUN="/"), 2,
               sqrt(sres$values),FUN="/")


# CONTRIBUTIONS TO FACTORS BY ROWS AND COLUMNS
# Contributions: mass times projection distance squared.
temp <- sweep( rproj^2, 1, fI, FUN="*")
# Normalize such that sum of contributions for a factor = 1.
sumCtrF <- apply(temp, 2, sum)
# NOTE: Obs. x factors.
# Read cntrs. with factors 1,2,... from cols. 2,3,...
rcntr <- sweep(temp, 2, sumCtrF, FUN="/")
temp <- sweep( cproj^2, 1, fJ, FUN="*")
sumCtrF <- apply(temp, 2, sum)
# NOTE: Vbs. x factors.
# Read cntrs. with factors 1,2,... from cols. 2,3,...
ccntr <- sweep(temp, 2, sumCtrF, FUN="/")


# CORRELATIONS WITH FACTORS BY ROWS AND COLUMNS
# dstsq(i) = sum_j 1/fj (fj^i - fj)^2
temp <- sweep(fJsupI, 2, fJ, "-")
dstsq <- apply( sweep( temp^2, 2, fJ, "/"), 1, sum)
# NOTE: Obs. x factors.
# Read corrs. with factors 1,2,... from cols. 2,3,...
rcorr <- sweep(rproj^2, 1, dstsq, FUN="/")
temp <- sweep(fIsupJ, 1, fI, "-")
dstsq <- apply( sweep( temp^2, 1, fI, "/"), 2, sum)
# NOTE: Vbs. x factors.
# Read corrs. with factors 1,2,... from cols. 2,3,...
ccorr <- sweep(cproj^2, 1, dstsq, "/")


# Value of this function on return: list containing
# projections, correlations, and contributions for rows
# (observations), and for columns (variables).
# In all cases, allow for first trivial first eigenvector.
list(rproj=rproj[,-1], rcorr=rcorr[,-1], rcntr=rcntr[,-1],
     cproj=cproj[,-1], ccorr=ccorr[,-1], ccntr=ccntr[,-1])


}
```

Some comments on this program are as follows. The program variables follow closely the notation to be more fully introduced and discussed in the next chapter. The following two paragraphs can be skipped at a first reading, and returned to later, following the reading of Chapter 2, in order to check the workings of the software code.

Firstly, the relative frequency table, f_{IJ}, that will be looked at further in the next chapter, is defined. So too are the marginals (vectors), f_I and f_J. They will be defined in the next chapter. Then we proceed to f_J^I and f_I^J. Next we have $s = f_J^I \circ f_{IJ}$, which yields a cross-product matrix on the J set of variables. By the time we get to $s2$ we have a matrix of term jj' equal to: $\sum_{i=1}^n f_{ij} f_{ij'} / f_i f_j'$. It is this matrix that we diagonalize. Informally we can say that this matrix is the equivalent in correspondence analysis to the correlation matrix in principal components analysis. A matrix-based description of the formulas here can be found in [50].

We determine the eigenvalues and eigenvectors. We take care of the fact that in correspondence analysis (due to linear dependence that is part and parcel of the weighting used, when we divide by f_I and f_J) the first eigenvalue is always 1. Therefore we ignore it. Note how a normalization term has to be applied to the eigenvectors.

In the next three paragraphs in the program we read off: (i) the projections on the newly found factors; (ii) the contributions to these factors; and (iii) the correlations with these factors.

1.4.4 Running the Analysis and Displaying Results

Putting it all together we have the following. We show the situation for a PC (Intel) computer running Windows. The situation would not be much difficult for a Unix computer. For example, sourcing the correspondence analysis program would be just `source("ca.r")`.

```
# Globular cluster data set read in:
x <- matrix(scan("c:/sodata.dat"), nrow=14, ncol=8, byrow=T)
source("c:/fuzzy.r")              # fuzzy coding of a vector
source("c:/ca.r")                 # correspondence analysis
source("c:/plaxes.r")             # draw axes
# In recoding in the following, we should of course write a
# simpler function!
y <- cbind(fuzzy(x[,1]),fuzzy(x[,2]),fuzzy(x[,3]),fuzzy(x[,4]),
           fuzzy(x[,5]),fuzzy(x[,6]),fuzzy(x[,7]),fuzzy(x[,8]))

# First the correspondence analysis.

xc <- ca(y)
# We find the eigenvalues to be the following:
# Eigenvalues follow (trivial first eigenvalue removed).
```

```
# 0.3937290 0.176932 0.1226037 0.05631895 0.04725499
# 0.01971215 0.00981572 0.004900663 0 0 0 0 0 0
# Eigenvalue rate, in thousandths.
# 473.6492 212.8461 147.4901 67.75071 56.84693 23.71337
# 11.80814 5.895412 0 0 0 0 0 0
lbls <- c("M15","M68","M13","M3","M5","M4","47 Tuc","M30",
        "NGC 6397", "M92","M12","NGC 6752","M10","M71")
clbls <- c("log(t_rel)","LOG(T_REL)","R_gc","R_GC","Z_g","Z_G",
        "log(mass)","LOG(MASS)","c","C","[Fe/H]","[FE/H]",
        "x","X","x_0","X_0")
# Plot an empty principal plane, setting up labels, and so on.
plot(xc$cproj[,1],xc$cproj[,2],xlab="Factor 1",ylab="Factor 2",
        type="n")
# Now plot the variable labels.
text(xc$cproj[,1],xc$cproj[,2],clbls)
plaxes(xc$cproj[,1],xc$cproj[,2])
# Now plot the row labels.
text(xc$rproj[,1],xc$rproj[,2],lbls,font=4)
title("Principal factor plane")


# Next the hierarchical clustering.
# We will use the already available R program (which we
#        originally wrote in Fortran in the 1980s).

# The variance of a pair of points motivates the use of 1/2
#        in the following.
dd <- 0.5*dist(xc$rproj)^2
xh <- hclust(dd, method="ward")
plclust(xh, labels=lbls, sub="", xlab="")
title
("Hier. clustering of the fuzzily (piecewise linear) coded data")
```

1.4.5   Hierarchical Clustering

Although the hierarchical clustering program in R has served us just fine here, we want on occasion to allow for weights. So let us look at this again, with the following program. Later – in Chapter 2 – we will look at C code that can be accessed in R, and which will provide for far greater computational efficiency and storage potential.

First, the way we use this alternative hierarchical clustering program is as follows.

```
source("c:/ahc.r")
unitwts <- rep(1,16)
xh <- hierclust(xc$rproj, unitwts)
```

```
plot(as.dendrogram(xh))
```

That is all. We only have sequence numbers here as labels. But for very large numbers of observations, this is acceptable.

The hierarchical clustering program uses two functions: one to calculate dissimilarities, dissim; and the second to find nearest neighbors from the dissimilarity array, getnns. Then the main program is called hierclust. Further details of this algorithm will be provided in the next chapter (section 2.5).

```
dissim <- function(a, wt) {
# Inputs.    a: matrix, for which we want distances on rows,
#            wt: masses of each row.
# Returns.   matrix of dims. nrow(a) x nrow(a) with
#            wtd. sqd. Eucl. distances.

    n <- nrow(a)
    m <- ncol(a)
    adiss <- matrix(0, n, n)

    for (i1 in 2:n) {
        adiss[i1,i1] <- 0.0
        for (i2 in 1:(i1-1)) {
            adiss[i1,i2] <- 0.0
            for (j in 1:m) {
                # We use the squared Euclidean distance, weighted.
                adiss[i1,i2] <- adiss[i1,i2] +
                    (wt[i1]*wt[i2])/(wt[i1]+wt[i2]) *
                    (a[i1,j]-a[i2,j])^2
            }
            adiss[i2,i1] <- adiss[i1,i2]
        }
    }
    adiss
}


getnns <- function(diss, flag) {
# Inputs.    diss: full distance matrix.
#            flag: "live" rows indicated by 1 are to be processed.
# Returns. List of: nn, nndiss.
#            nn:    list of nearest neighbor of each row.
#            nndiss: nearest neighbor distance of each row.

    nn <- rep(0, nrow(diss))
```

```
    nndiss <- rep(0.0, nrow(diss))
    MAXVAL <- 1.0e12
    if (nrow(diss) != ncol(diss))
        stop("Invalid input first parameter.")
    if (nrow(diss) != length(flag))
        stop("Invalid inputs 1st/2nd parameters.")

    for (i1 in 1:nrow(diss)) {
        if (flag[i1] == 1) {
            minobs <- -1
            mindis <- MAXVAL
            for (i2 in 1:ncol(diss)) {
                if ( (diss[i1,i2] < mindis) && (i1 != i2) ) {
                    mindis <- diss[i1,i2]
                    minobs <- i2
                }
            }
            nn[i1] <- minobs
            nndiss[i1] <- mindis
        }
    }
    list(nn = nn, nndiss = nndiss)
}

hierclust <- function(a, wt) {

    MAXVAL <- 1.0e12

    n <- nrow(a)
    diss <- dissim(a, wt)              # call to function dissim
    flag <- rep(1, n)                  # active/dead indicator
    a <- rep(0, n-1)                   # left subnode on clustering
    b <- rep(0, n-1)                   # right subnode on clustering
    ia <- rep(0, n-1)                  # R-compatible version of a
    ib <- rep(0, n-1)                  # R-compatible version of b
    lev <- rep(0, n-1)                 # level or criterion values
    card <- rep(1, n)                  # cardinalities
    mass <- wt
    order <- rep(0, n)                 # R-compatible order for plotting

    nnsnnsdiss <- getnns(diss, flag) # call to function getnns
    clusmat <- matrix(0, n, n)       # cluster memberships
    for (i in 1:n) clusmat[i,n] <- i # init. trivial partition

    for (ncl in (n-1):1) {             # main loop
```

```
# check for agglomerable pair
minobs <- -1;
mindis <- MAXVAL;
for (i in 1:n) {
    if (flag[i] == 1) {
        if (nnsnnsdiss$nndiss[i] < mindis) {
            mindis <- nnsnnsdiss$nndiss[i]
            minobs <- i
        }
    }
}
# find agglomerands clus1 and clus2, with former < latter
if (minobs < nnsnnsdiss$nn[minobs]) {
    clus1 <- minobs
    clus2 <- nnsnnsdiss$nn[minobs]
}
if (minobs > nnsnnsdiss$nn[minobs]) {
    clus2 <- minobs
    clus1 <- nnsnnsdiss$nn[minobs]
}
# Agglomeration of pair clus1 < clus2 defines cluster ncl

#--------------------------- Block for subnode labels
a[ncl] <- clus1                  # aine, or left child node
b[ncl] <- clus2                  # benjamin, or right child node
# Now build up ia, ib as version of a, b which is R-compliant
if (card[clus1] == 1) ia[ncl] <- (-clus1)       # singleton
if (card[clus2] == 1) ib[ncl] <- (-clus2)       # singleton
if (card[clus1] > 1) {           # left child is non-singleton
    lastind <- 0
    for (i2 in (n-1):(ncl+1)) { # Must have n-1 >= ncl+1 here
        if (a[i2] == clus1) lastind <- i2  # Only concerns a[i2]
    }
    ia[ncl] <- n - lastind       # label of non-singleton
}
if (card[clus2] > 1) {           # right child is non-singleton
    lastind <- 0
    for (i2 in (n-1):(ncl+1)) { # Must have n-1 >= ncl+1 here
        if (a[i2] == clus2) lastind <- i2  # Only concerns a[i2]
    }
    ib[ncl] <- n - lastind       # label of non-singleton
}
if (ia[ncl] > 0 || ib[ncl] > 0) { # Check that left < right
    left <- min(ia[ncl],ib[ncl])
    right <- max(ia[ncl],ib[ncl])
```

```
        ia[ncl] <- left                    # Just get left < right
        ib[ncl] <- right
    }
    #-------------------------------------------------------

    lev[ncl] <- mindis
    for (i in 1:n) {
        clusmat[i,ncl] <- clusmat[i,ncl+1]
        if (clusmat[i,ncl] == clus2) clusmat[i,ncl] <- clus1
    }
    # Next we need to update diss array
    for (i in 1:n) {
        if ( (i != clus1) && (i != clus2) && (flag[i] == 1) ) {
            diss[clus1,i] <-
              ((mass[clus1]+mass[i])/
              (mass[clus1]+mass[clus2]+mass[i]))*diss[clus1,i] +
              ((mass[clus2]+mass[i])/
              (mass[clus1]+mass[clus2]+mass[i]))*diss[clus2,i] -
              (mass[i]/
              (mass[clus1]+mass[clus2]+mass[i]))*diss[clus1,clus2]
            diss[i,clus1] <- diss[clus1,i]
        }
    }
    # Update mass of new cluster
    mass[clus1] <- mass[clus1] + mass[clus2]

    # Update card of new cluster
    card[clus1] <- card[clus1] + card[clus2]

    # Cluster label clus2 is knocked out;
    # following is not necessary but is done for rigor
    flag[clus2] <- 0
    nnsnnsdiss$nndiss[clus2] <- MAXVAL
    mass[clus2] <- 0.0
    for (i in 1:n) {
        diss[clus2,i] <- MAXVAL
        diss[i,clus2] <- diss[clus2,i]
    }
    # Finally update nnsnnsdiss$nn and nnsnnsdiss$nndiss
    # i.e. nearest neighbors and the nearest neigh. dissimilarity
    nnsnnsdiss <- getnns(diss, flag)
}

temp <- cbind(a,b)
merge2 <- temp[nrow(temp):1, ]
```

```
      temp <- cbind(ia,ib)
      merge <- temp[nrow(temp):1,]
      dimnames(merge) <- NULL
      # merge is R-compliant; merge2 is an alternative format

      #----------------------- Build R-compatible order from ia, ib
      orderlist <- c(merge[n-1,1], merge[n-1,2])
      norderlist <- 2
      for (i in 1:(n-2)) { # For precisely n-2 further node expansions
          for (i2 in 1:norderlist) {    # Scan orderlist
              if (orderlist[i2] > 0) { # Non-singleton to be expanded
                  tobeexp <- orderlist[i2]
                  if (i2 == 1) {
                      orderlist <- c(merge[tobeexp,1],merge[tobeexp,2],
                                     orderlist[2:norderlist])
                  }
                  if (i2 == norderlist) {
                      orderlist <- c(orderlist[1:(norderlist-1)],
                                     merge[tobeexp,1],merge[tobeexp,2])
                  }
                  if (i2 > 1 && i2 < norderlist) {
                      orderlist <- c(orderlist[1:(i2-1)],
                                     merge[tobeexp,1],merge[tobeexp,2],
                                     orderlist[(i2+1):norderlist])
                  }
                  norderlist <- length(orderlist)
              }
          }
      }
      orderlist <- (-orderlist)
      class(orderlist) <- "integer"
      xcall <- "hierclust(a,wt)"
      class(xcall) <- "call"

      retlist <- list(merge=merge,height=lev[(n-1):1],
          order=orderlist,labels=dimnames(a)[[1]],
          method="minvar",call=xcall,
          dist.method="euclidean-factor",clusmat=clusmat)
      class(retlist) <- "hclust"
      retlist
}
```

1.4.6 Handling Large Data Sets

The computationally critical part of the correspondence analysis program is the eigenvalue and eigenvector calculation. This is $O(m^3)$ where m is the number of columns in the matrix that is input to this eigen-reduction. Say that n, the number of rows in the input data table, X, is from a few dozen in number up to 100 or 200. Say that m, the number of columns in the input data table, is a few dozen.

Then if data X is read into R variable x, the following command is fine: xc <- ca(x).

However if m, the number of columns, is a few hundred, the eigen-reduction will take a very long time. In this case one can simply transpose the data, and then of course take care that row and column projections, correlations and contributions, are all reversed (rows become columns and vice versa): xc <- ca(t(x)).

The hierarchical clustering program discussed above has computational complexity of the order of n^2 where n is the number of rows. R handles badly the iterative operations needed here. This is because any interpreted language like R (interpreted meaning that commands can be executed on the fly, one command after the other; the alternative is a compiled language like C, Java, Fortran or Pascal, where efficient binary code is distilled in a first, compilation phase, from the complete program) is inherently inefficient for loop operations.

The best solution in this case is to use a compiled language like C, and link that into the R session. In the next chapter we will discuss a hierarchical clustering program in C, for use in the R environment, which can be used for large values of n (many thousands).

2

Theory of Correspondence Analysis

This chapter will give a solid grounding in the theory underlying correspondence analysis and associated analysis algorithms – hierarchical clustering in particular. This chapter is not essential for reading later chapters with examples and discussion of software. It can safely be used for reference, on points of detail, when and where the need arises.

2.1 Vectors and Projections

An $n \times m$ input data array may be viewed as a set of n row-vectors, or alternatively as a set of m column-vectors. In Figure 2.1 (left), three points are located in $\mathbb{R}^2$. On axis 1 the points are fairly regularly laid out, with coordinates 1, 2 and 3, whereas on axis 2 it appears that the points with projections 4 and 5 are somewhat separated from the point with projection 2. In higher dimensional spaces we are limited to being easily able to visualize one-dimensional and two-dimensional representations (axes and planes), although with more difficulty we can construct a three-dimensional representation.

Given, for example, the array of 4 objects by 5 attributes,

$$\begin{pmatrix} 7 & 3 & 4 & 1 & 6 \\ 3 & 4 & 7 & 2 & 0 \\ 1 & 7 & 3 & -1 & 4 \\ 2 & 0 & -6 & 4 & 1 \end{pmatrix}$$

the projection of the 4 objects onto the plane constituted by axes 1 and 3 is simply

$$\begin{pmatrix} 7 & 4 \\ 3 & 7 \\ 1 & 3 \\ 2 & -6 \end{pmatrix}$$

Thus far, the projection of any point onto an axis or plane is the trivial operation of reading off the projection value from a data table. Principal components analysis and correspondence analysis, however, obtain *better* axes. Consider Figure 2.1 (right) where a new axis has been drawn as nearly as

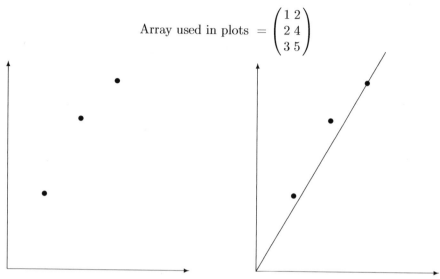

Array used in plots $= \begin{pmatrix} 1 & 2 \\ 2 & 4 \\ 3 & 5 \end{pmatrix}$

FIGURE 2.1

Points and their projections onto axes. The three points have coordinates (1,2), (2,4) and (3,5). So projections on axis 1 (horizontal axis) are 1, 2 and 3; and projections on axis 2 (vertical axis) are 2, 4 and 5. On the right, a "best fitting" line is drawn through the three points.

possible through all points. It is clear that if this axis went precisely through all points, then a second axis would be redundant in defining the locations of the points; i.e., the cloud of three points would be seen to be one-dimensional.

Eigen-reduction seeks the axes for which the cloud of points are closest, with usually the Euclidean distance defining *closeness*. This criterion is identical to another criterion: that the projections of points on any axis sought be as elongated as possible. This second criterion is that the *variance* of the projections be as great as possible. Variance is defined as average distance squared. When mass or weight is considered, one speaks of inertia, and it is this latter term that is used in correspondence analysis.

In general the points under examination are m-dimensional, and it is rare to find that they approximately lie on a one-dimensional surface, i.e., a line. A second best-fitting axis, orthogonal to the first already found, will together constitute a best-fitting plane. Then a third best-fitting axis, orthogonal to the two already obtained, will collectively constitute a best-fitting three-dimensional subspace.

A few simple examples in two-dimensional space are shown in Figure 2.2.

Consider the case where the points are *centered* (i.e., the origin is located at the center of gravity). We will seek the best-fitting axis, and then the next best-fitting axis. Figure 2.2a consists of just two points, which if centered

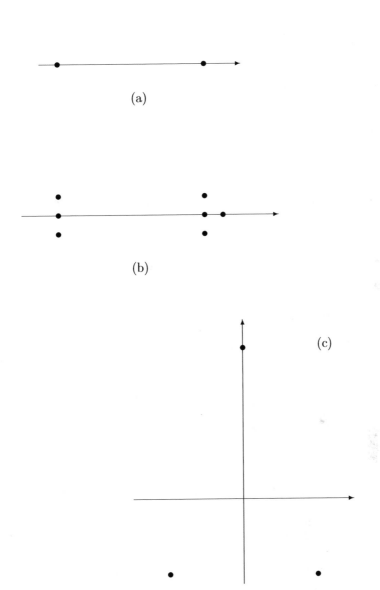

FIGURE 2.2
Some examples of principal components analysis of centered clouds of points.

must lie on a one-dimensional axis. In Figure 2.2c, the points are arranged at the vertices of a triangle. The vertical axis, here, accounts for the greatest variance, and the symmetry of the problem necessitates the positioning of this axis as shown. In the examples of Figure 2.2, and generally, we find that the positive and negative orientations of the axes could be changed without any loss of information or generality.

2.2   Factors

2.2.1   Review of Metric Spaces

Let E be a vector space of dimension n: for example $\mathbb{R}^n$. Let f be some symmetric positive definite bilinear form:

$$f : E \times E \longrightarrow \mathbb{R}$$

$$(x, y) \longrightarrow f(x, y) \quad \text{for } x, y \in E$$

The mapping f is symmetric (i.e., $f(x, y) = f(y, x)$), definite (i.e., $f(x, x) = 0 \iff x = 0$), positive (i.e. $x \neq y \iff f(x, x) > 0$ and bilinear (i.e., $f(\lambda x, y) = f(x, \lambda y) = \lambda f(x, y)$; $f(x, y_1 + y_2) = f(x, y_1) + f(x, y_2)$).

The function f defines: (i) a scalar product: $\langle x, y \rangle = \langle y, x \rangle = x'y = xy' = f(x, y)$, the scalar product of x and y, where prime denotes transpose; (ii) a Euclidean norm: $\|x\|^2 = f(x, x)$; (iii) a Euclidean distance: $d(x, y) = \|x - y\|$; and (iv) orthogonality: $f(x, y) = 0 \iff x$ is f-orthogonal to y.

A Euclidean space is therefore defined either through a symmetric positive definite bilinear form f, or its associated Euclidean distance d. A quadratic form is a mapping taking $x \in E$ into $f(x, x)$. A norm is an example of a quadratic form.

If we apply a linear mapping given by matrix M to the vector space E, in the transform space we can use this same linear mapping to define scalar product, distance, norm and orthogonality using the analogous function g: $g(x, y) = x'My$. To satisfy the requirements of g being symmetric, positive and definite, we require M to be a symmetric positive definite matrix. We then can define the norm ($\|x\|_M^2 = x'Mx$), the Euclidean distance ($d_M(x, y) = \|x - y\|_M$) and M-orthogonality ($\langle x, y \rangle_M = x'My = 0$ if x is M-orthogonal to y).

The classical scalar product and Euclidean distance are associated with $M = I_n$, the identity matrix. A normalization often applied is to require each coordinate to have unit variance: therefore elements of M off the principal diagonal are 0, and the ith term on the principal diagonal is $1/\sigma_i^2$, the inverse of the variance. A more expressive normalization can be applied where M is the inverse of the variance-covariance matrix: the resulting modified Euclidean

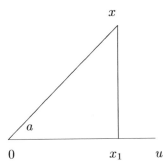

FIGURE 2.3

Projection of x at x_1 on axis u.

distance is termed the Mahalanobis distance. We see in these examples how the use of a symmetric, positive definite matrix, M, can generalize the usual Euclidean distance. One can speak of the Euclidean metric (or distance) associated with M, or simply the Euclidean metric M.

Consider the case of centered n-valued coordinates or variables, x_i. The sum of variable vectors is a constant, proportional to the mean variable. The upshot of this is that the centered vectors lie on a hyperplane, or a sub-space, of dimension $n - 1$. Consider a probability distribution p defined on I, i.e., for all i we have $p_i > 0$ (note: strictly greater than 0 to avoid the temporary inconvenience of a lower dimensional subspace) and $\sum_{i \in I} p_i = 1$. We will often use the set I of indices, $i, i \in I$, to represent the point set. Let M_{p_I} be the diagonal matrix with diagonal elements consisting of the p terms. This matrix is the covariance metric. We have: $x' M_{p_I} x = \sum_{i \in I} p_i x_i^2 = \text{var}(x)$; and $x' M_{p_I} y = \sum_{i \in I} p_i x_i y_i = \text{cov}(x, y)$.

Consider the mapping f_x of $y \in E$ onto $f(x, y) \in \mathbb{R}$. Since f above was termed a bilinear form, f_x will be termed a linear form. Cailliez and Pagès [30] consider the mapping of x into f_x as a mapping (an isomorphism) from a space into a dual space. The data matrix to be analyzed in correspondence analysis is an example of such a linear form.

Scalar product defines orthogonal projection. It is denoted, as already seen: $x'y$, or $\langle x, y \rangle$. Therefore the (orthogonal) projection of the vector or n-dimensional point x on the vector u is $x'Mu$ (Figure 2.3). The projected value, ordinarily referred to as the projection, or coordinate, is $x_1 = (x'Mu/u'Mu)$. Here $x_1 u$ and u are both vectors, and of course the scalar products yield scalars. Let us consider the norm of vector $x_1 u$: this is $(x'Mu/u'Mu)\|u\|$. This can be simplified to: $(x'Mu/\|u\|)$, whenever u is taken as of unit norm in the M metric. The quantity $(x'Mu)/(\|x\|\|u\|)$ can be interpreted as the cosine of the angle a between vectors x and u.

We looked at the covariance metric above. The points of I, we noted, were centered. Therefore adding the point-vectors yields a zero-vector result:

the center is at the origin. The centered points of I lie on the subspace of codimension 1, or alternatively in a subspace of dimension $n-1$. Let each diagonal element of the diagonal matrix M_{p_I} be the probability p_i, as we have also seen above. With the Euclidean covariance norm, associated with the restriction of M_{p_I} to the hyperplane of the centered variables, is associated in the dual space the χ^2 norm of center p_I associated with the restriction of M_{1/p_I} to the hyperplane of measures of zero total mass.

Let us pause briefly to explain some of the terms used here. The space of observations, and the space of variables, are very often termed dual spaces. The "restriction to the hyperplane of centered variables" simply means that the variables (often the columns of the data table) are centered. The "restriction to the hyperplane of measures of zero total mass" simply means that the center – in this case, the center of the observations, or origin of the space – is by design zero-valued on all coordinates.

Consider probability density r_I. Having $\sum_{i \in I}(r_i - q_i) = 0$ implies that $r_I - q_I \in G$ where G is the "restriction to the hyperplane of measures of zero total mass." The χ^2 distance relative to the center p_I between the probability densities r_I and q_I is given by:

$$\|r_I - q_I\|_{p_I}^2 = \langle (r_I - q_I), (r_I - q_I) \rangle_{p_I} = (r_I - q_I)' M_{1/p_I}(r_I - q_I) = \sum_{i \in I}(r_i - q_i)^2/p_i$$

$$(2.1)$$

The amount of information in a data table may be measured by its dependence on its marginal densities, using the squared χ^2 distance. Consider data tables p_{IJ} (of typical, ijth entry, p_{ij}) and $q_{IJ} = p_I p_J$ (of typical, ijth entry, $p_i p_j$). The χ^2 distance of center q_{IJ} between the densities p_{IJ} and q_{IJ} is

$$\|p_{IJ} - q_{IJ}\|_{q_{IJ}}^2 = \sum_{(i,j) \in I \times J}(p_{ij} - p_i p_j)^2/p_i p_j \tag{2.2}$$

Within a multiplicative function of n, this is exactly the quantity which can be assessed with a χ^2 test with $n-1$ degrees of freedom.

The χ^2 distance is much used in correspondence analysis. Under appropriate circumstances (when $p_I = p_J = $ constant) then it becomes a classical Euclidean distance.

2.2.2   Clouds of Points, Masses, and Inertia

We now look at the basic operations of correspondence analysis in a way that is close to the software implementation. In particular we use I to denote the labels or indices of the observation set; and J to denote the labels or indices of the attribute set.

The moment of inertia of a cloud of points in a Euclidean space, with both distances and masses defined, is the sum for all elements of I of the products of mass by distance squared from the center of the cloud:

$$M^2(N_J(I)) = \sum_{i \in I} f_i \|f_J^i - f_J\|_{f_J}^2 = \sum_{i \in I} f_i \rho^2(i) \qquad (2.3)$$

In the latter term, ρ is the Euclidean distance from the cloud center, and f_i is the mass of element i. The mass is the marginal distribution of the input data table. Let us take a step back: the given contingency table data is denoted $k_{IJ} = \{k_{IJ}(i,j) = k(i,j); i \in I, j \in J\}$. We have $k(i) = \sum_{j \in J} k(i,j)$. Analogously $k(j)$ is defined, and $k = \sum_{i \in I, j \in J} k(i,j)$. Next, $f_{IJ} = \{f_{ij} = k(i,j)/k; i \in I, j \in J\} \subset \mathbb{R}_{I \times J}$, similarly f_I is defined as $\{f_i = k(i)/k; i \in I, j \in J\} \subset \mathbb{R}_I$, and f_J analogously.

Next back to the first right hand side term in equation 2.3: the conditional distribution of f_J knowing $i \in I$, also termed the jth profile with coordinates indexed by the elements of I, is

$$f_J^i = \{f_j^i = f_{ij}/f_i = (k_{ij}/k)/(k_i/k); f_i \neq 0; j \in J\}$$

and likewise for f_I^j.

The cloud of points consists of the couple: profile coordinate and mass. We have $N_J(I) = \{(f_J^i, f_i); j \in J\} \subset \mathbb{R}_J$, and again similarly for $N_I(J)$.

From equation 2.3, it can be shown that

$$M^2(N_J(I)) = M^2(N_I(J)) = \|f_{IJ} - f_I f_J\|_{f_I f_J}^2 = \sum_{i \in I, j \in J} (f_{ij} - f_i f_j)^2 / f_i f_j$$

$$(2.4)$$

The term $\|f_{IJ} - f_I f_J\|_{f_I f_J}^2$ is the χ^2 metric between the probability distribution f_{IJ} and the product of marginal distributions $f_I f_J$, with as center of the metric the product $f_I f_J$.

In correspondence analysis, the choice of χ^2 metric of center f_J is linked to the *principle of distributional equivalence*, explained as follows. Consider two elements j_1 and j_2 of J with identical profiles: i.e., $f_I^{j_1} = f_I^{j_2}$. Consider now that elements (or columns) j_1 and j_2 are replaced with a new element j_s such that the new coordinates are aggregated profiles, $f_{ij_s} = f_{ij_1} + f_{ij_2}$, and the new masses are similarly aggregated: $f_{ij_s} = f_{ij_1} + f_{ij_2}$. Then there is *no effect* on the distribution of distances between elements of I. The distance between elements of J, other than j_1 and j_2, is naturally not modified. This description has followed closely [47] (chapter 2).

The principle of distributional equivalence leads to representational self-similarity: aggregation of rows or columns, as defined above, leads to the same analysis. Therefore it is very appropriate to analyze a contingency table with fine granularity, and seek in the analysis to merge rows or columns, through aggregation.

2.2.3 Notation for Factors

Correspondence analysis produces an ordered sequence of pairs, called factors, (F_α, G_α) associated with real numbers called eigenvalues $0 \leq \lambda_\alpha \leq 1$. The number of eigenvalues and associated factor couples is: $\alpha = 1, 2, \ldots, N = \min(|\,I\,| - 1, |\,J\,| - 1)$, where $|\,.\,|$ denotes set cardinality. We denote $F_\alpha(I)$ the set of values of the factor of rank α for elements i of I; and similarly $G_\alpha(J)$ denotes the values of the factor of rank α for all elements j of J. We see that F is a function on I, and G is a function on J.

2.2.4 Properties of Factors

$\sum_{i \in I} f_i F_\alpha(i) = 0; \quad \sum_{j \in J} f_j G_\alpha(j) = 0$ (centered).

$\sum_{i \in I} f_i F_\alpha^2(i) = \lambda_\alpha; \quad \sum_{j \in J} f_j G_\alpha^2(j) = \lambda_\alpha$ (moments of inertia).

$\sum_{i \in I} f_i F_\alpha(i) F_\beta(i) = \delta_{\alpha\beta}$ (orthonormality).

$\sum_{j \in J} f_j G_\alpha(j) G_\beta(j) = \delta_{\alpha\beta}$ (orthonormality).

The Dirac delta notation used here expresses the following: $\delta_{\alpha\beta} = 0$ if $\alpha \neq \beta$ and $= 1$ if $\alpha = \beta$.

Normalized factors: on the sets I and J, we next define the functions ϕ^I and ψ^J of zero mean, of unit variance, pairwise uncorrelated on I (respectively J), and associated with masses f_J (respectively f_I).

$\sum_{i \in I} f_i \phi_\alpha(i) = 0; \quad \sum_{j \in J} f_j \psi_\alpha(j) = 0$

$\sum_{i \in I} f_i \phi_\alpha^2(i) = 1; \quad \sum_{j \in J} f_j \psi_\alpha^2(j) = 1$

$\sum_{i \in I} f_i \phi_\alpha(i) \phi_\beta(i) = \delta_{\alpha\beta}; \quad \sum_{j \in J} f_j \psi_\alpha(j) \psi_\beta(j) = \delta_{\alpha\beta}$

Between unnormalized and normalized factors, we have the relations:

$\phi_\alpha(i) = \lambda_\alpha^{-\frac{1}{2}} F_\alpha(i) \;\; \forall i \in I, \;\; \forall \alpha = 1, 2, \ldots N$

$\psi_\alpha(j) = \lambda_\alpha^{-\frac{1}{2}} G_\alpha(j) \;\; \forall j \in J, \;\; \forall \alpha = 1, 2, \ldots N$

The moment of inertia of the clouds $N_J(I)$ and $N_I(J)$ in the direction of the α axis is λ_α.

2.2.5 Properties of Factors: Tensor Notation

We consider a tensor calculus of transitions between probability spaces. A transition from I to J is an element of the tensor product $\mathbb{R}_J \otimes \mathbb{R}^I$. It is a function on I, but with values in the J measures; or the conditional probability of j given i. Such a transition takes masses (or probability measures or densities) from I to J; and associates every function on J with a function on I.

We write: $f_J^I \in \mathbb{R}_J \otimes \mathbb{R}^I$, signifying that f_J^I is a measure relative to I, and a function relative to J. In turn this implies that $f_J^i \in \mathbb{R}_J$ is a measure on J, and is a function of $i \in I$.

Analogously the transition of J to I is written $f_I^J \in \mathbb{R}_I \otimes \mathbb{R}^J$.

Taking a step back in the data we analyze, the normalized frequencies viewed as probabilities are given by $f_{IJ} \in \mathbb{R}_I \otimes \mathbb{R}_J$.

The marginals or masses, which are measures on I and J, can be written:

$f_I = \delta_I^{IJ} \circ f_{IJ}$ and $f_J = \delta_J^{IJ} \circ f_{IJ}$

where δ is the Kronecker (or discrete Dirac) symbol.

In this notation we can also write: $f_{IJ} = (\delta_I^I \times f_J^I) \circ f_I$, and $f_{IJ} = (f_I^J \times \delta_J^J) \circ f_J$.

Composition of measure and function transitions will be illustrated with the definitions and operations for factors. A transition of I to J can be used to transport any measure μ_I (on I) in order to obtain a measure μ_J on J; and in the other direction it can be used to associate with every function G^J (on J) a function F^I (on I).

Thus, we have: $\pi_J = f_J^I \circ \mu_I$ which, in detail, is: $\pi_j = \{\pi_j | j \in J\}$ and $\forall j \in J : \pi_j = \sum_{i \in I} f_j^i \mu_i$.

For the factors, then: $F^I = G^J \circ f_J^I$, and in greater detail: $F^I = \{F^i | i \in I\}$ and $\forall i \in I : F^i = \sum_{j \in J} G^j f_j^i$.

Composition rules (up and down index combination) can be noted here. These tensor composition rules are an extended form of product conformability for matrices.

Quadratic form of the moments of inertia, relative to the origin, of the cloud $N(J)$:

$$\sigma_{II} = (f_I^J \cdot f_I^J) \circ f_J$$

$$\sigma_{ii'} = \sum_j f_i^j f_{i'}^j f_j = \sum_j (f_{ij} f_{i'j}/f_j)$$

It can be shown ([15], p. 153) that the principal eigenvalue λ corresponding to eigenvector ϕ satisfies: $\phi^I \circ f_I^J \circ f_J^I - (\phi^I \circ f_I)\delta^I = \lambda \phi^I$. Furthermore it holds that $\delta^I \circ f_I^J \circ f_J^I = (\delta^I \circ f_K)\delta^I = \delta^I$; that is to say, δ^I is the first trivial eigenvector, i.e., the constant function equal to 1. The factor ϕ^I is zero mean for the measure f_I, i.e., $\phi^I \circ f_I = 0$.

Since $\phi^I \circ (f_J^I \circ f_I^J) = \lambda \phi^I$, we can right-multiply by f_I^J to get $(\phi^I \circ f_J^I) \circ (f_J^I \circ f_I^J) = \lambda(\phi^I \circ f_I^J)$. Consequently $\phi^I \circ f_I^J$ is a factor of the dual space.

Through consideration of the norms, it turns out that we can define factors on J, ϕ^J, in the following way: $\phi^J = (1/\sqrt{\lambda(\phi)})\phi^I \circ f_I^J$.

Benzécri [15] argues in favor of tensor notation: firstly to take account of more than two arguments or indices; and secondly to render symmetries much clearer than would otherwise be possible. Further motivation can be added: a matrix expresses a linear mapping (of rows onto columns or vice versa), a linear mapping of a set into itself, a bilinear form on the cross-product of a set with itself, and so on; with tensor notation, these different cases are clearly distinguished.

2.3 Transform

2.3.1 Forward Transform

We have seen that the χ^2 metric is defined in direct space, i.e., the space of profiles. The Euclidean metric is defined for the factors. We can characterize correspondence analysis as the mapping of a cloud in χ^2 space to Euclidean space. (Later, we will characterize hierarchical agglomerative clustering as the mapping of any metric or indeed non-metric space into an ultrametric space.)

Distances between profiles are as follows.

$$\|f_J^i - f_J^{i'}\|_{f_J}^2 = \sum_{j\in J} \left(f_j^i - f_j^{i'}\right)^2 / f_j = \sum_{\alpha=1..N} (F_\alpha(i) - F_\alpha(i'))^2$$

$$\|f_I^j - f_I^{j'}\|_{f_I}^2 = \sum_{i\in I} \left(f_i^j - f_i^{j'}\right)^2 / f_i = \sum_{\alpha=1..N} (G_\alpha(j) - G_\alpha(j'))^2 \quad (2.5)$$

Norm, or distance of a point $i \in N_J(I)$ from the origin or center of gravity of the cloud $N_J(I)$, is:

$$\rho^2(i) = \|f_J^i - f_J\|_{f_J}^2 = \sum_{\alpha=1..N} F_\alpha^2(i)\rho^2(j) = \|f_I^j - f_I\|_{f_I}^2 = \sum_{\alpha=1..N} F_\alpha^2(j) \quad (2.6)$$

2.3.2 Inverse Transform

The correspondence analysis transform, taking profiles into a factor space, is reversed with no loss of information as follows $\forall (i,j) \in I \times J$.

$$f_{ij} = f_i f_j \left(1 + \sum_{\alpha=1..N} \lambda_\alpha^{-\frac{1}{2}} F_\alpha(i)G_\alpha(j)\right)$$

For profiles we have:

$$f_i^j = f_i \left(1 + \sum_\alpha \lambda_\alpha^{-\frac{1}{2}} F_\alpha(i)G_\alpha(j)\right); \quad f_j^i = f_j \left(1 + \sum_\alpha \lambda_\alpha^{-\frac{1}{2}} F_\alpha(i)G_\alpha(j)\right)$$

2.3.3 Decomposition of Inertia

We know (equation 2.6) that the distance of a point from the center of gravity of the cloud is

$$\rho^2(i) = \|f_J^i - f_J\|_{f_J}^2 = \sum_{j\in J} \left(f_j^i - f_j\right)^2 / f_j$$

Decomposition of the cloud's inertia is:

$$M^2(N_J(I)) = \sum_{\alpha=1..N} \lambda_\alpha = \sum_{i\in I} f_i \rho^2(i) \qquad (2.7)$$

In greater detail, we have for this decomposition:

$$\lambda_\alpha = \sum_{i\in I} f_i F_\alpha^2(i) \text{ and } \rho^2(i) = \sum_{\alpha=1..N} F_\alpha^2(i)$$

2.3.4 Relative and Absolute Contributions

$f_i \rho^{(i)}$ is the absolute contribution of point i to the inertia of the cloud, $M^2(N_J(I))$, or the variance of point i.
$f_i F_\alpha^2(i)$ is the absolute contribution of point i to the moment of inertia λ_α.
$f_i F_\alpha^2(i)/\lambda_\alpha$ is the relative contribution of point i to the moment of inertia λ_α. (Often denoted CTR.)
$F_\alpha^2(i)$ is the contribution of point i to the χ^2 distance between i and the center of the cloud $N_J(I)$.
$\cos^2 a = F_\alpha^2(i)/\rho^2(i)$ is the relative contribution of the factor α to point i. (Often denoted COR.)
Based on the latter term, we have: $\sum_{\alpha=1..N} F_\alpha^2(i)/\rho^2(i) = 1$.
Analogous formulas hold for the points j in the cloud $N_I(J)$.

2.3.5 Reduction of Dimensionality

Interpretation is usually limited to the first few factors.

Decomposition of inertia is usually far less decisive than (cumulative) percentage variance explained in principal components analysis. One reason for this is that, in correspondence analysis, often recoding tends to bring input data coordinates closer to vertices of a hypercube. This is especially the case with boolean forms of coding, as used in complete disjunctive form.

The term QLT is an overall quality of representation in the factor space. $QLT(i) = \sum_{\alpha=1..N'} \cos^2 a$, where angle a has been defined above (previous subsection, also denoted COR) and where $N' < N$ is the quality of representation of element i in the factor space of dimension N'.

INR represents overall inertia. $INR(i) = \rho^2(i)$, defined for all i, is the distance squared in factor space of element i from the center of gravity of the cloud.

WTS represents the set of weights or masses. $WTS(i) = f_i$, for all i, is the mass or marginal frequency of the element i.

2.3.6 Interpretation of Results

The mains steps in interpreting results of correspondence analysis are often the following.

1. Projections onto factors 1 and 2, 2 and 3, 1 and 3, etc. of set I, set J, or both sets simultaneously.

2. Spectrum of non-increasing values of eigenvalues.

3. Interpretation of axes. We can distinguish between the general (latent semantic, conceptual) meaning of axes, and axes which have something specific to say about groups of elements. Usually contrast is important: what is found to be analogous at one extremity versus the other extremity; or oppositions or polarities.

4. Factors are determined by how much the elements contribute to their dispersion. Therefore the values of CTR are examined in order to identify or to name the factors (for example, with higher order concepts). (Informally, CTR allows us to work from the elements towards the factors.)

5. The values of COR are squared cosines, which can be considered as being like correlation coefficients. If COR(i, α) is large (say, around 0.8) then we can say that that element is well explained by the axis of rank α. (Informally, COR allows us to work from the factors towards the elements.)

2.3.7   Analysis of the Dual Spaces

We have:

$$F_\alpha(i) = \lambda_\alpha^{-\frac{1}{2}} \sum_{j \in J} f_j^i G_\alpha(j) \text{ for } \alpha = 1, 2, \ldots N; i \in I$$

$$G_\alpha(j) = \lambda_\alpha^{-\frac{1}{2}} \sum_{i \in I} f_i^j F_\alpha(i) \text{ for } \alpha = 1, 2, \ldots N; j \in J$$

These are termed the *transition formulas*. The coordinate of element $i \in I$ is the barycenter of the coordinates of the elements $j \in J$, with associated masses of value given by the coordinates of f_j^i of the profile f_j^i. This is all to within the $\lambda_\alpha^{-\frac{1}{2}}$ constant.

We also have:

$$\phi_\alpha(i) = \lambda_\alpha^{-\frac{1}{2}} \sum_{j \in J} f_j^i \psi_\alpha(j)$$

and

$$\psi_\alpha(j) = \lambda_\alpha^{-\frac{1}{2}} \sum_{i \in I} f_i^j \phi_\alpha(i)$$

This implies that we can pass easily from one space to the other, i.e., we carry out the diagonalization, or eigen-reduction, in the more computationally favorable space which is usually $\mathbb{R}^J$. In the output display, the barycentric principle comes into play: this allows us to simultaneously view and interpret observations and attributes.

2.3.8 Supplementary Elements

Overly-preponderant elements (i.e., row or column profiles), or exceptional elements (e.g., a sex attribute, given other performance or behavioral attributes) may be placed as supplementary elements. This means that they are given zero mass in the analysis, and their projections are determined using the transition formulas. This amounts to carrying out a correspondence analysis first, without these elements, and then projecting them into the factor space following the determination of all properties of this space.

2.4 Algebraic Perspective

2.4.1 Processing

From the initial frequencies data matrix, a set of probability data, x_{ij}, is defined by dividing each value by the grand total of all elements in the matrix. In correspondence analysis, each row (or column) point is considered to have an associated weight. The weight of the ith row point is given by $x_i = \sum_j x_{ij}$, and the weight of the jth column point is given by $x_j = \sum_i x_{ij}$. We consider the row points to have coordinates x_{ij}/x_i, thus allowing points of the same *profile* to be identical (i.e., superimposed). The following weighted Euclidean distance, the χ^2 distance, is then used between row points:

$$d^2(i,k) = \sum_j \frac{1}{x_j} \left(\frac{x_{ij}}{x_i} - \frac{x_{kj}}{x_k} \right)^2$$

and an analogous distance is used between column points.

The mean row point is given by the weighted average of all row points:

$$\sum_i x_i \frac{x_{ij}}{x_i} = x_j$$

for $j = 1, 2, \ldots, m$. Similarly the mean column profile has ith coordinate x_i.

2.4.2 Motivation

Correspondence analysis is not unlike principal components analysis in its underlying geometrical bases. While principal components analysis is particularly suitable for quantitative data, correspondence analysis is appropriate for the following types of input data: frequencies, contingency tables, probabilities, categorical data, and mixed qualitative/categorical data.

In the case of *frequencies* (i.e., the ijth table entry indicates the frequency of occurrence of attribute j for object i) the row and column "profiles" are of interest. That is to say, the relative magnitudes are of importance. Use of a weighted Euclidean distance, termed the χ^2 distance, gives a zero distance for example to the following 5-coordinate vectors which have identical profiles of values: (2,7,0,3,1) and (8,28,0,12,4). Probability type values can be constructed here by dividing each value in the vectors by the sum of the respective vector values.

A particular type of frequency of occurrence data is the contingency table, a table crossing sets of characteristics of the population under study. As an example, an $n \times m$ contingency table might give frequencies of the existence of n different metals in stars of m different ages. Correspondence analysis allows the study of the two sets of variables which constitute the rows and columns of the contingency table. In its usual variant, principal components analysis would give priority to either the rows or the columns by standardizing: if, however, we are dealing with a contingency table, both rows and columns are equally interesting. The "standardizing" inherent in correspondence analysis (a consequence of the χ^2 distance) treats rows and columns in an identical manner. One byproduct is that the row and column projections in the new space may both be plotted on the same output graphic presentations.

Categorical data may be coded by the "scoring" of 1 (presence) or 0 (absence) for each of the possible categories. Such coding leads to *complete disjunctive coding*. It will be noted below how correspondence analysis of an array of such complete disjunctive data is referred to as multiple correspondence analysis, and how such a coding of categorical data is, in fact, closely related to contingency table type data.

2.4.3   Operations

As in the case of principal components analysis, we first consider the projections of the n profiles in $\mathbb{R}^m$ onto an axis, $\mathbf{u}$. This is given by

$$\sum_j \frac{x_{ij}}{x_i} \frac{1}{x_j} u_j$$

for all i (note the use of the scalar product here). Let the above, for convenience, be denoted by w_i.

The weighted sum of projections uses weights x_i (i.e., the row masses), since the inertia of projections is to be maximized. Hence the quantity to be maximized is

$$\sum_i x_i w_i^2$$

subject to the vector **u** being of unit length (this, as in principal components analysis, is required since otherwise vector **u** could be taken as unboundedly large):

$$\sum_j \frac{1}{x_j} u_j{}^2 = 1$$

It may then be verified using Lagrangian multipliers that optimal **u** is an eigenvector of the matrix of dimensions $m \times m$ whose (j, k)th term is

$$\sum_i \frac{x_{ij} x_{ik}}{x_i x_k}$$

where $1 \leq j, k \leq m$. (This matrix is not symmetric, and a related symmetric matrix must be constructed for eigen-reduction: we will not detail this here.) The associated eigenvalue, λ, indicates the importance of the best fitting axis, or eigenvalue: it may be expressed as the *percentage of inertia explained* relative to subsequent, less good fitting, axes.

The results of a correspondence analysis are centered (x_j and x_i are the jth and ith coordinates, or the average profiles, of the origin of the output graphic representations). The first eigenvalue resulting from correspondence analysis is a trivial one, of value 1; the associated eigenvector is a vector of 1s [50, 85].

2.4.4 Axes and Factors

In the previous subsection it has been seen that projections of points onto axis **u** were with respect to the $1/x_i$ weighted Euclidean metric. This makes interpreting projections very difficult from a human/visual point of view, and so it is more natural to present results in such a way that projections can be simply appreciated. Therefore *factors* are defined, such that the projections of row vectors onto factor ϕ associated with axis **u** are given by

$$\sum_j \frac{x_{ij}}{x_i} \phi_j$$

for all i. Taking

$$\phi_j = \frac{1}{x_j} u_j$$

ensures this and projections onto ϕ are with respect to the ordinary (unweighted) Euclidean distance.

An analogous set of relationships hold in $\mathbb{R}^n$ where the best fitting axis, **v**, is searched for. A simple mathematical relationship holds between **u** and **v**, and between ϕ and ψ (the latter being the factor associated with eigenvector **v**):

$$\sqrt{\lambda} \psi_i = \sum_j \frac{x_{ij}}{x_i} \phi_j$$

$$\sqrt{\lambda}\phi_j = \sum_i \frac{x_{ij}}{x_j}\psi_i$$

These are termed *transition formulas*. Axes $\mathbf{u}$ and $\mathbf{v}$, and factors ϕ and ψ, are associated with eigenvalue λ and best fitting higher-dimensional subspaces are associated with decreasing values of λ, determined in the diagonalization.

The transition formulas allow *supplementary rows* or columns to be projected into either space. If ξ_j is the jth element of a supplementary row, with mass ξ, then a factor loading is simply obtained subsequent to the correspondence analysis:

$$\psi_i = \frac{1}{\sqrt{\lambda}} \sum_j \frac{\xi_j}{\xi}\phi_j$$

A similar formula holds for supplementary columns. Such supplementary elements are therefore "passive" and are incorporated into the correspondence analysis results subsequent to the eigen-analysis being carried out.

2.4.5 Multiple Correspondence Analysis

When the input data is in complete disjunctive form, correspondence analysis is termed multiple correspondence analysis. Complete disjunctive form is a form of coding where the response categories, or modalities, of an attribute have one and only one non-zero response (see Figure 2.4a). Ordinarily correspondence analysis is used for the analysis of contingency tables: such a table may be derived from a table in complete disjunctive form by taking the matrix product between its transpose and itself. The symmetric table obtained in this way is referred to as a *Burt table*. Correspondence analysis of either table gives similar results, only the eigenvalues differing [50, 85].

A few features of the analysis of tables in complete disjunctive form will be mentioned.

- The modalities (or response categories) of each attribute in multiple correspondence analysis have their center of gravity at the origin.

- The number of nonzero eigenvalues found is less than or equal to the total number of modalities less the total number of attributes.

- Due to this large dimensionality of the space being analyzed, it is not surprising that eigenvalues tend to be very small in multiple correspondence analysis. It is not unusual to find that the first few factors can be usefully interpreted and yet account for only a few percent of the total inertia.

The principal steps in interpreting the output of multiple correspondence analysis are as for the case of correspondence analysis.

- The Burt table is scanned for significantly high frequencies of co-occurrence.

Type			Age			Properties				
T1	T2	T3	A1	A2	A3	P1	P2	P3	P4	P5
1	0	0	0	1	0	0	0	0	0	1
0	1	0	0	0	1	0	0	0	0	1
1	0	0	0	0	1	0	0	1	0	0
1	0	0	0	1	0	0	0	1	0	0
1	0	0	1	0	0	1	0	0	0	0
0	0	1	0	1	0	1	0	0	0	0
0	0	1	1	0	0	0	0	0	0	1

(a) Table in complete disjunctive form.

	T1	T2	T3	A1	A2	A3	P1	P2	P3	P4	P5
T1	4	0	0	1	2	1	1	0	2	0	1
T2	0	1	0	0	0	1	0	0	0	0	1
T3	0	0	2	1	1	0	1	0	0	0	1
A1	1	0	1	2	0	0	1	0	0	0	1
A2	2	0	1	0	3	0	1	0	1	0	1
A3	1	1	0	0	0	2	0	0	1	0	1
P1	1	0	1	1	1	0	2	0	0	0	0
P2	0	0	0	0	0	0	0	0	0	0	0
P3	2	0	0	0	1	1	0	0	2	0	0
P4	0	0	0	0	0	0	0	0	0	0	0
P5	1	1	1	1	1	1	0	0	0	0	3

(b) Burt table.

Notes:

- Attributes: type, age, properties.

- Modalities: T1, T2, ..., P5.

- Row sums of table in complete disjunctive form are constant, and equal to the number of attributes.

- Each attribute × attribute submatrix of the Burt table (e.g., ages × ages) is necessarily diagonal, with column totals of the table in complete disjunctive form making up the diagonal values.

FIGURE 2.4
Table in complete disjunctive form and associated Burt table.

- The axes are interpreted in order of decreasing importance using the modalities which contribute most, in terms of inertia, to the axes (i.e., mass times projected distance squared). The projection coordinates serve to indicate how far the modality can be assessed relative to the axis.

- The planar graphic representations (projections of row and column points in the plane formed by factors 1 and 2, and by other pairs of factors) are examined.

- The interrelationships between modalities, relative to the axes, are examined, and substantive conclusions are drawn.

It may be noted that in correspondence analysis as looked at in this section, the coding used is such that the row-points are of constant weight. This allows, quite generally, user intervention in the weighting of rows relative to columns. In our experience, we have often obtained similar results for a principal components analysis with the usual standardization to zero mean and unit standard deviation, on the one hand; and on the other, a correspondence analysis with twice the number of columns as the matrix analyzed by principal components analysis such that for each column j we also have a column j' with value $x_{ij'} = \max_k x_{ik} - x_{ij}$. This is referred to as *doubling* the data.

Some typical output configurations can arise, the most well known being the "horseshoe" shaped curve associated with pronounced linearity in the data. Figure 2.5 gives an example of the type of doubled data for which this pattern arises. It may be explained by the constraints imposed on the pairwise distances resulting from the input data. The one-dimensional ordering inherent in the input data is referred to as a *seriation*.

2.4.6 Summary of Correspondence Analysis Properties

See Table 2.1.

2.5 Clustering

2.5.1 Hierarchical Agglomerative Clustering

Hierarchical agglomeration on n observation vectors, $i \in I$, involves a series of $1, 2, \ldots, n-1$ pairwise agglomerations of observations or clusters, with the following properties. A hierarchy $H = \{q | q \in 2^I\}$ such that (i) $I \in H$, (ii) $i \in H \ \forall i$, and (iii) for each $q \in H, q' \in H : q \cap q' \neq \emptyset \Longrightarrow q \subset q'$ or $q' \subset q$. An indexed hierarchy is the pair (H, ν) where the positive function defined on H, i.e., $\nu : H \to \mathbb{R}^+$, satisfies: $\nu(i) = 0$ if $i \in H$ is a singleton; and

1	1	1	1	1	0	0	0	0	0	0	0	0	0	0
2	0	0	1	1	1	1	0	0	0	0	0	0	0	0
3	0	0	0	0	1	1	1	1	0	0	0	0	0	0
4	0	0	0	0	0	0	1	1	1	1	0	0	0	0
5	0	0	0	0	0	0	0	0	1	1	1	1	0	0
6	0	0	0	0	0	0	0	0	0	0	1	1	1	1

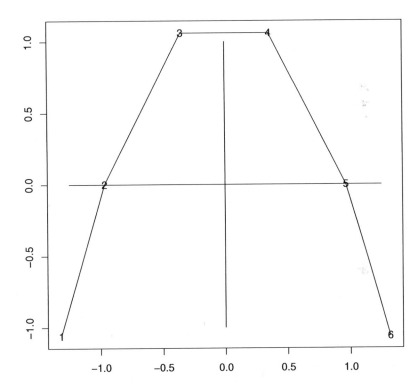

FIGURE 2.5
Horseshoe pattern in principal plane of correspondence analysis.

TABLE 2.1

Properties of observation (row) space $\mathbb{R}^m$ and attribute (column) space $\mathbb{R}^n$ in correspondence analysis.

Space $\mathbb{R}^m$

1. n row points, each of m coordinates.

2. The jth coordinate is x_{ij}/x_i.

3. The mass of point i is x_i.

4. The χ^2 distance between row points i and k is:
$$d^2(i,k) = \sum_j \frac{1}{x_j} \left(\frac{x_{ij}}{x_i} - \frac{x_{kj}}{x_k} \right)^2.$$
Hence this is a Euclidean distance, with respect to the weighting $1/x_j$ (for all j), between *profile* values x_{ij}/x_i, etc.

5. The criterion to be optimized: the weighted sum of squares of projections, where the weighting is given by x_i (for all i).

Space $\mathbb{R}^n$

1. m column points, each of n coordinates.

2. The ith coordinate is x_{ij}/x_j.

3. The mass of point j is x_j.

4. The χ^2 distance between column points g and j is:
$$d^2(g,j) = \sum_i \frac{1}{x_i} \left(\frac{x_{ig}}{x_g} - \frac{x_{ij}}{x_j} \right)^2.$$
Hence this is a Euclidean distance, with respect to the weighting $1/x_i$ (for all i), between *profile* values x_{ig}/x_g, etc.

5. The criterion to be optimized: the weighted sum of squares of projections, where the weighting is given by x_j (for all j).

(ii) $q \subset q' \implies \nu(q) < \nu(q')$. Function ν is the agglomeration level. Take $q \subset q'$, let $q \subset q''$ and $q' \subset q''$, and let q'' be the lowest level cluster for which this is true. Then if we define $D(q, q') = \nu(q'')$, D is an ultrametric. In practice, we start with a Euclidean or other dissimilarity, use some criterion such as minimizing the change in variance resulting from the agglomerations, and then define $\nu(q)$ as the dissimilarity associated with the agglomeration carried out.

2.5.2 Minimum Variance Agglomerative Criterion

For Euclidean distance inputs, the following definitions hold for the minimum variance or Ward error sum of squares agglomerative criterion:

- Coordinates of the new cluster center, following agglomeration of q and q', where m_q is the mass of cluster q defined as cluster cardinality, and (vector) q denotes using overloaded notation the center of (set) cluster q: $q'' = (m_q q + m_{q'} q')/(m_q + m_{q'})$.

- Following the agglomeration of q and q', we define the following dissimilarity: $(m_q m_{q'})/(m_q + m_{q'})\|q - q'\|^2$.

These two definitions are all we need to specify the hierarchical clustering algorithm. When q and q' are both singletons, the latter rule implies that a weighting of 0.5 is applied to the Euclidean distance. Hierarchical clustering based on factor projections, if desired using a limited number of factors (e.g., 7) in order to filter out the most useful information in our data, provides for a consistent framework. In such a case, hierarchical clustering can be seen to be a mapping of Euclidean distances into ultrametric distances.

2.5.3 Lance-Williams Dissimilarity Update Formula

Hierarchic agglomerative algorithms may be conveniently broken down into two groups of methods. The first group is that of linkage methods – the single, complete, weighted and unweighted average linkage methods. These are methods for which a graph representation can be used. The second group of hierarchic clustering methods are methods which allow the cluster centers to be specified (as an average or a weighted average of the member vectors of the cluster). These methods include the centroid, median and minimum variance methods. The latter may be specified either in terms of dissimilarities, alone or alternatively in terms of cluster center coordinates and dissimilarities. A very convenient formulation, in dissimilarity terms, which embraces all the hierarchical methods mentioned so far, is the *Lance-Williams dissimilarity update formula*. If points (objects) i and j are agglomerated into cluster $i \cup j$, then we must simply specify the new dissimilarity between the cluster and all other points (objects or clusters). The formula is:

$$d(i \cup j, k) = \alpha_i d(i, k) + \alpha_j d(j, k) + \beta d(i, j) + \gamma \mid d(i, k) - d(j, k) \mid$$

where α_i, α_j, β and γ define the agglomerative criterion. Values of these are listed in the second column of Table 2.2.

In the case of the single link method, using $\alpha_i = \alpha_j = \frac{1}{2}$, $\beta = 0$ and $\gamma = -\frac{1}{2}$ gives us

$$d(i \cup j, k) = \frac{1}{2}d(i, k) + \frac{1}{2}d(j, k) - \frac{1}{2} \mid d(i, k) - d(j, k) \mid$$

which, it may be verified by taking a few simple examples of three points, i, j and k, can be rewritten as

$$d(i \cup j, k) = \min \{d(i, k), d(j, k)\}$$

In the case of the methods which use cluster centers, we have the center coordinates (in column 3 of Table 2.2) and dissimilarities as defined between cluster centers (column 4 of Table 2.2). The Euclidean distance must be used, initially, for equivalence between the two approaches. In the case of the *median method*, for instance, we have the following (cf. Table 2.2).

Let $\mathbf{a}$ and $\mathbf{b}$ be two points (i.e., m-dimensional vectors: these are objects or cluster centers) which have been agglomerated, and let $\mathbf{c}$ be another point. From the Lance-Williams dissimilarity update formula, using squared Euclidean distances, we have:

$$
\begin{aligned}
d^2(a \cup b, c) &= \frac{d^2(a,c)}{2} + \frac{d^2(b,c)}{2} - \frac{d^2(a,b)}{4} \\
&= \frac{\|\mathbf{a}-\mathbf{c}\|^2}{2} + \frac{\|\mathbf{b}-\mathbf{c}\|^2}{2} - \frac{\|\mathbf{a}-\mathbf{b}\|^2}{4}
\end{aligned}
\tag{2.8}
$$

The new cluster center is $(\mathbf{a} + \mathbf{b})/2$, so that its distance to point $\mathbf{c}$ is

$$\|\mathbf{c} - \frac{\mathbf{a}+\mathbf{b}}{2}\|^2 \tag{2.9}$$

That these two expressions are identical is readily verified. The correspondence between these two perspectives on the one agglomerative criterion is similarly proved for the centroid and minimum variance methods.

The single linkage algorithm, duly generalized for the use of the Lance-Williams dissimilarity update formula, is applicable for all agglomerative strategies. The appropriate update formula listed in Table 2.2 is used in step 2 of the algorithm that follows. For cluster center methods, and with suitable alterations for graph methods, the following algorithm is an alternative to the general dissimilarity based algorithm (the latter may be described as a "stored dissimilarities approach").

Stored data approach

Step 1 Examine all interpoint dissimilarities, and form cluster from two closest points.

TABLE 2.2

Specifications of seven hierarchical clustering methods. Notes: $|\,i\,|$ is the number of objects in cluster i; $\mathbf{g}_i$ is a vector in m-space (m is the set of attributes), – either an initial point or a cluster center; $\|.\|$ is the norm in the Euclidean metric; the names UPGMA, etc. are due to Sneath and Sokal [80]; finally, the Lance and Williams recurrence formula is:

$$d_{i \cup j, k} = \alpha_i d_{ik} + \alpha_j d_{jk} + \beta d_{ij} + \gamma \,|\, d_{ik} - d_{jk} \,|$$

Hierarchical clustering methods (and aliases)	Lance and Williams dissimilarity update formula	Coordinates of center of cluster, which agglomerates clusters i and j	Dissimilarity between cluster centers g_i and g_j
Single link (nearest neighbor)	$\alpha_i = 0.5$ $\beta = 0$ $\gamma = -0.5$ (More simply: $min\{d_{ik}, d_{jk}\}$)		
Complete link (diameter)	$\alpha_i = 0.5$ $\beta = 0$ $\gamma = 0.5$ (More simply: $max\{d_{ik}, d_{jk}\}$)		
Group average (average link, UPGMA)	$\alpha_i = \frac{|i|}{|i|+|j|}$ $\beta = 0$ $\gamma = 0$		
McQuitty's method (WPGMA)	$\alpha_i = 0.5$ $\beta = 0$ $\gamma = 0$		
Median method (Gower's, WPGMC)	$\alpha_i = 0.5$ $\beta = -0.25$ $\gamma = 0$	$\mathbf{g} = \frac{\mathbf{g}_i + \mathbf{g}_j}{2}$	$\|\mathbf{g}_i - \mathbf{g}_j\|^2$
Centroid (UPGMC)	$\alpha_i = \frac{|i|}{|i|+|j|}$ $\beta = -\frac{|i||j|}{(|i|+|j|)^2}$ $\gamma = 0$	$\mathbf{g} = \frac{|i|\mathbf{g}_i + |j|\mathbf{g}_j}{|i|+|j|}$	$\|\mathbf{g}_i - \mathbf{g}_j\|^2$
Ward's method (minimum variance, error sum of squares)	$\alpha_i = \frac{|i|+|k|}{|i|+|j|+|k|}$ $\beta = -\frac{|k|}{|i|+|j|+|k|}$ $\gamma = 0$	$\mathbf{g} = \frac{|i|\mathbf{g}_i + |j|\mathbf{g}_j}{|i|+|j|}$	$\frac{|i||j|}{|i|+|j|}\|\mathbf{g}_i - \mathbf{g}_j\|^2$

Step 2 Replace two points clustered by the representative point (center of gravity) or by cluster fragment.

Step 3 Return to step 1, treating clusters as well as remaining objects, until all objects are in one cluster.

In steps 1 and 2, "point" refers either to objects or clusters, both of which are defined as vectors in the case of cluster center methods. This algorithm is justified by storage considerations, since we have $O(n)$ storage required for n initial objects and $O(n)$ storage for the $n-1$ (at most) clusters. In the case of linkage methods, the term "fragment" in step 2 refers (in the terminology of graph theory) to a connected component in the case of the single link method and to a clique or complete subgraph in the case of the complete link method. The overall complexity of the above algorithm is $O(n^3)$: the repeated calculation of dissimilarities in step 1, coupled with $O(n)$ iterations through steps 1, 2 and 3. Note however that this does not take into consideration the extra processing required in a linkage method, where "closest" in step 1 is defined with respect to graph fragments.

2.5.4 Reciprocal Nearest Neighbors and Reducibility

Some very efficient improvements on both the stored data, and stored dissimilarity, algorithms have been proposed (for a survey, see [60, 62]). In particular there is the *nearest neighbor (NN) chain* algorithm. Here is a short description of this approach.

A *NN*-chain consists of an arbitrary point (a in Figure 2.6); followed by its *NN* (b in Figure 2.6); followed by the *NN* from among the remaining points (c, d and e in Figure 2.6) of this second point; and so on until we necessarily have some pair of points which can be termed reciprocal or mutual *NN*s. (Such a pair of *RNN*s may be the first two points in the chain; and we have assumed that no two dissimilarities are equal.)

In constructing a *NN*-chain, irrespective of the starting point, we may agglomerate a pair of *RNN*s as soon as they are found. What guarantees that we can arrive at the same hierarchy as if we used the "stored dissimilarities" or "stored data" algorithms described earlier in this section? Essentially this is the same condition as that under which no inversions or reversals are produced by the clustering method. Figure 2.7 gives an example of this, where d is agglomerated at a lower criterion value (i.e., dissimilarity) than was the case at the previous agglomeration.

This is formulated as:

Inversion impossible if: $d(i,j) < d(i,k)$ or $d(j,k) \Rightarrow d(i,j) < d(i \cup j, k)$

a   b   c   d   e

FIGURE 2.6
Five points, showing *NN*s and *RNN*s.

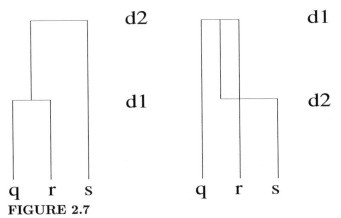

FIGURE 2.7
Alternative representations of a hierarchy with an inversion.

2.5.5 Nearest-Neighbor Chain Algorithm

Using the Lance-Williams dissimilarity update formula, it can be shown that the minimum variance method does not give rise to inversions; neither do the linkage methods; but the median and centroid methods cannot be guaranteed not to have inversions.

To return to Figure 2.6, if we are dealing with a clustering criterion which precludes inversions, then c and d can justifiably be agglomerated, since no other point (for example, b or e) could have been agglomerated to either of these.

The processing required, following an agglomeration, is to update the *NN*s of points such as b in Figure 2.6 (and on account of such points, this algorithm was initially dubbed *algorithme des célibataires*, the bachelor/spinster algorithm). The following is a summary of the algorithm:

NN-chain algorithm

Step 1 Select a point arbitrarily.

Step 2 Grow the *NN*-chain from this point until a pair of *RNN*s are obtained.

Step 3 Agglomerate these points (replacing with a cluster point, or updating the dissimilarity matrix).

Step 4 From the point which preceded the *RNN*s (or from any other arbitrary point if the first two points chosen in steps 1 and 2 constituted a pair of *RNN*s), return to step 2 until only one point remains.

2.5.6 Minimum Variance Method in Perspective

The search for clusters of maximum homogeneity leads to the minimum variance criterion. Since no coordinate axis is prioritized by the Euclidean distance, the resulting clusters will be approximately hyperspherical. Such ball-shaped clusters will therefore be very unsuitable for examining straggly patterns of points. However, in the absence of information about such patterns in the data, homogeneous clusters will provide the most useful condensation of the data.

The following properties make the minimum variance agglomerative strategy particularly suitable for synoptic clustering:

1. As discussed in the section to follow, the two properties of cluster homogeneity and cluster separability are incorporated in the cluster criterion. For summarizing data, it is unlikely that more suitable criteria could be devised.

2. As in the case of other geometric strategies, the minimum variance method defines a cluster center of gravity. This mean set of cluster members' coordinate values is the most useful summary of the cluster. It may also be used for the fast selection and retrieval of data, by matching on these cluster representative vectors rather than on each individual object vector.

3. A top-down hierarchy traversal algorithm may also be implemented for information retrieval. Using a query vector, the left or right subtree is selected at each node for continuation of the traversal (it is best to ensure that each node has precisely two successor nodes in the construction of the hierarchy). Such an algorithm will work best if all top-down traversals through the hierarchy are of approximately equal length. This will be the case if and only if the hierarchy is as "symmetric" or "balanced" as possible (see Figure 2.8). Such a balanced hierarchy is usually

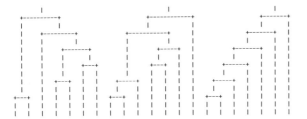

FIGURE 2.8
Three binary hierarchies: balanced, unbalanced and intermediate.

of greatest interest for interpretative purposes also: a partition, derived from a hierarchy, and consisting of a large number of small classes, and one or a few large classes, is less likely to be of practical use.

For such reasons, a "symmetric" hierarchy is desirable. It has been shown, using a number of different measures of hierarchic symmetry, that the minimum variance, closely followed by the complete link, methods generally give the most symmetric hierarchies [61].

4. Unlike other geometric agglomerative methods – in particular the centroid and the median methods (see definitions, Table 2.2, above) – the sequence of agglomerations in the minimum variance method is guaranteed not to allow inversions in the cluster criterion value. Inversions or reversals (Figure 2.7) are inconvenient, and can make interpretation of the hierarchy difficult.

5. Finally, computational performance has in the past favored linkage based agglomerative criteria, and in particular the single linkage method. The computational advances described above for the minimum variance method (principally the *NN*-chain algorithm) make it increasingly attractive for practical applications involving large amounts of data.

2.5.7 Minimum Variance Method: Mathematical Properties

The minimum variance method produces clusters which satisfy compactness and isolation criteria. These criteria are incorporated into the dissimilarity, noted in Table 2.2, as will now be shown.

In Ward's method, we seek to agglomerate two clusters, c_1 and c_2, into cluster c such that the within-class variance of the partition thereby obtained is minimum. Alternatively, the between-class variance of the partition obtained is to be maximized. Let P and Q be the partitions prior to, and subsequent to, the agglomeration; let p_1, p_2, ... be classes of the partitions:

$$P = \{p_1, p_2, \ldots, p_k, c_1, c_2\}$$
$$Q = \{p_1, p_2, \ldots, p_k, c\}$$

As usual, I is the set of such objects, and i denotes any individual or object. In the following, classes (i.e., p or c) and individuals (i.e., i) will be considered as vectors or as sets: the context, and the block typing of vectors, will be sufficient to make clear which is the case.

Total variance of the cloud of objects in m-dimensional space is decomposed into the sum of within-class variance and between-class variance. This is Huyghen's theorem in classical mechanics. Let V denote *variance*. The total variance of the cloud of objects is

$$V(I) = \frac{1}{n} \sum_{i \in I} (\mathbf{i} - \mathbf{g})^2$$

where $\mathbf{g}$ is the grand mean of the n objects: $\mathbf{g} = \frac{1}{n} \sum_{i \in I} \mathbf{i}$. The between-class variance is

$$V(P) = \sum_{p \in P} \frac{|p|}{n} (\mathbf{p} - \mathbf{g})^2$$

where $|p|$ is the cardinality of (i.e. number of members in) class p. (Note that $\mathbf{p}$ – in block type-face – is used to denote the center of gravity – a vector – and p the set whose center of gravity this is). Finally, the within-class variance is

$$\frac{1}{n} \sum_{p \in P} \sum_{i \in p} (\mathbf{i} - \mathbf{p})^2$$

For two partitions, before and after an agglomeration, we have respectively:

$$V(I) = V(P) + \sum_{p \in P} V(p)$$

$$V(I) = V(Q) + \sum_{p \in Q} V(p)$$

Hence,

$$V(P) + V(p_1) + \ldots + V(p_k) + V(c_1) + V(c_2)$$
$$= V(Q) + V(p_1) + \ldots + V(p_k) + V(c)$$

Therefore:

$$V(Q) = V(P) + V(c_1) + V(c_2) - V(c)$$

In agglomerating two classes of P, the variance of the resulting partition (i.e., $V(Q)$) will necessarily decrease: therefore in seeking to minimize this decrease, we simultaneously achieve a partition with maximum between-class variance. The criterion to be optimized can then be shown to be:

$$V(P) - V(Q) = V(c) - V(c_1) - V(c_2)$$
$$= \frac{|c_1|\,|c_2|}{|c_1|+|c_2|}\|\mathbf{c_1} - \mathbf{c_2}\|^2$$

which is the dissimilarity given in Table 2.2. This is a dissimilarity which may be determined for any pair of classes of partition P; and the agglomerands are those classes, c_1 and c_2, for which it is minimum.

It may be noted that if c_1 and c_2 are singleton classes, then $V(\{c_1, c_2\}) = \frac{1}{2}\|\mathbf{c_1} - \mathbf{c_2}\|^2$ (i.e., the variance of a pair of objects is equal to half their Euclidean distance).

2.5.8 Simultaneous Analysis of Factors and Clusters

The barycentric principle (see section 2.3.7, "Analysis of the dual spaces") allows both row points and column points to be displayed simultaneously as projections. We therefore can consider the following outputs for a correspondence analysis: (i) simultaneous display of I and J; (ii) tree on I; and (iii) tree on J. To help analyze these outputs we can explore the representation of clusters (derived from the hierarchical trees) in factor space, leading to programs traditionally called FACOR; and the representation of clusters in the profile coordinate space, leading to programs traditionally called VACOR.

In the case of FACOR, for every couple q, q' of a partition of I, we calculate

$$\frac{(f_q f'_q)}{(f_q + f'_q)}\|q - q'\|^2$$

This can be decomposed using the axes of $\mathbb{R}_J$, as well as using the factorial axes.

In the case of VACOR, we can explore the cluster dipoles which take account of the "elder" and "younger" cluster components. See Figure 2.9.

We have $F_\alpha(a) = \sum_{i \in q}(f_i/f_q)F_\alpha(i)$. We consider the vectors defining the dipole: $[q, a(q)]$ and $[q, b(q)]$. We then study the squared cosine of the angle between vector $[a(q), b(q)]$ and the factorial axis of rank α. This squared cosine defines the relative contribution of the pair q, α to the level index $\nu(q)$ of the class q.

2.6 Questions

A number of questions are listed here, to help in further study. All of these questions can be answered from the material in this chapter.

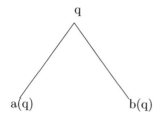

FIGURE 2.9
Node q has subnodes $a(q)$ and $b(a)$. The latter are dipoles of the cluster associated with q.

1. Write down the expressions for: (i) Euclidean distance, and (ii) χ^2 distance. Describe what conditions are needed so that the χ^2 distance becomes a Euclidean distance.

2. Explain, with equations, how row points are positioned in the output factor space at the center of gravity of all column points; and how column points are positioned at the center of gravity of row points.

3. Define inertia (or moment of inertia). Distinguish between: projections, contributions, and correlations.

4. How does multiple correspondence analysis differ from correspondence analysis in its more simple expression?

5. Explain the principle of operation of supplementary elements. When and where are they used?

6. Describe and explain hierarchical clustering, using the minimum variance criterion, and using a nearest neighbor algorithm, in the two cases of (i) starting with a data matrix, and (ii) starting with a matrix of distances.

2.7 Further R Software for Correspondence Analysis

Software discussed here is available on the web at: www.correspondances.info

2.7.1 Supplementary Elements

Supplementary rows, and supplementary columns, are projected retrospectively into the analysis. Here are a few reasons for their use. The term element is commonly used for either row or column.

- An element may be very different in nature from the other elements. Consider, for example, a gender or age attribute, when the data concerns online browsing or web purchasing behavior. Putting such an element as supplementary then allows associations between attributes and respondents to be found, but does not otherwise influence the analysis.

- An element may dominate and by putting it as supplementary we can assess its genuine significance.

- An element may be untrustworthy, in that it contains missing values, or uncertain values.

R code for caSuppRow.r, supporting supplementary rows, follows. Usage indication can be found at the start. This program will be used in later chapters.

```
caSuppRow <- function(xtab, rsupp) {

# Projections of supplementary rows.  One or more supplementary
# rows can be used.  Example of use:
# x <- read.table("c:/mandible77s.dat")          # Read data
# xca <- ca(x[1:77,])                            # Corr. anal.
# xcar <- caSuppRow(x[1:77,], x[78:86,])         # Suppl. rows
# plot(c(xca$rproj[,1],xca$cproj[,1]),           # Prepare plot
#      c(xca$rproj[,2],xca$cproj[,2]),
#      type="n", xlab="Factor 1 (50.0% of inertia)",
#      ylab="Factor 2 (21.0% of inertia)")
# text(xca$rproj[,1],xca$rproj[,2],dimnames(x)[[1]]) # Plot prin.rows
# text(xca$cproj[,1],xca$cproj[,2],dimnames(x)[[2]],font=4) #Plotcols
# text(xcar[,1],xcar[,2],dimnames(xcar)[[1]],font=3) # Plot supp.rows
# title("77 mandibles, 9 suppl. (groups), crossed by 9 attributes")

tot <- sum(xtab)
fIJ <- xtab/tot
fI <- apply(fIJ, 1, sum)
fJ <- apply(fIJ, 2, sum)
fJsupI <- sweep(fIJ, 1, fI, FUN="/")
fIsupJ <- sweep(fIJ, 2, fJ, FUN="/")
s <- as.matrix(t(fJsupI)) %*% as.matrix(fIJ)
s1 <- sweep(s, 1, sqrt(fJ), FUN="/")
s2 <- sweep(s1, 2, sqrt(fJ), FUN="/")
sres <- eigen(s2)
sres$values[sres$values < 1.0e-8] <- 0.0
# cat("Eigenvalues follow (trivial first eigenvalue removed).\n")
```

```
# cat(sres$values[-1], "\n")
# cat("Eigenvalue rate, in thousandths.\n")
# tot <- sum(sres$values[-1])
# cat(1000*sres$values[-1]/tot,"\n")
# Eigenvectors divided rowwise by sqrt(fJ):
evectors <- sweep(sres$vectors, 1, sqrt(fJ), FUN="/")
# rproj <- as.matrix(fJsupI) %*% evectors
temp    <- as.matrix(s2) %*% sres$vectors
# Following divides rowwise by sqrt(fJ) and
# columnwise by sqrt(eigenvalues):
# Note: first column of cproj is trivially 1-valued.
cproj <- sweep ( sweep(temp,1,sqrt(fJ),FUN="/"), 2,
                   sqrt(sres$values),FUN="/")


# Note: we must coerce rsupp to matrix type, which
# propagates to rsuppIJ
rsuppIJ <- as.matrix(rsupp)/tot
if (nrow(rsuppIJ) > 1) rsuppI <- apply(rsuppIJ, 1, sum)
if (nrow(rsuppIJ) == 1) rsuppI <- sum(rsuppIJ)


rsuppproj <- rsuppIJ %*% cproj
temp <- rsuppproj
# Divide cols. by mass; and then rows. by sqrt of evals.
rsuppproj <- sweep ( sweep(temp,1,rsuppI,FUN="/"),2,
                   sqrt(sres$values),FUN="/")


# Value of this function on return: table of projections,
# rows = set of supplementary rows; columns = set of factors.
# (More than 1 supplementary row => labels will be retained.)
# Adjust for trivial factor.
rsuppproj[,-1]


}
```

R code for supplementary columns, `caSuppCol.r`, follows. Usage indication can be found at the start. This program will be used in later chapters.

```
caSuppCol <- function(xtab, csupp) {

# Projections of supplementary columns.  One or more
# supplementary columns can be used.
# Example of use on goblets data (25 x 6):
# suppl <- caSuppCol(goblets[,-3], goblets[,3])
# This uses all except column 3 as principal; and
# column 3 as supplementary.
```

```
tot <- sum(xtab)
fIJ <- xtab/tot
fI <- apply(fIJ, 1, sum)
fJ <- apply(fIJ, 2, sum)
fJsupI <- sweep(fIJ, 1, fI, FUN="/")
fIsupJ <- sweep(fIJ, 2, fJ, FUN="/")
s <- as.matrix(t(fJsupI)) %*% as.matrix(fIJ)
s1 <- sweep(s, 1, sqrt(fJ), FUN="/")
s2 <- sweep(s1, 2, sqrt(fJ), FUN="/")
sres <- eigen(s2)
sres$values[sres$values < 1.0e-8] <- 0.0
# cat("Eigenvalues follow (trivial first eigenvalue removed).\n")
# cat(sres$values[-1], "\n")
# cat("Eigenvalue rate, in thousandths.\n")
# tot <- sum(sres$values[-1])
# cat(1000*sres$values[-1]/tot,"\n")
# Eigenvectors divided rowwise by sqrt(fJ):
evectors <- sweep(sres$vectors, 1, sqrt(fJ), FUN="/")
rproj <- as.matrix(fJsupI) %*% evectors

# Note: we must coerce csupp to matrix type, which
# propagates to csuppIJ
csuppIJ <- as.matrix(csupp)/tot
if (ncol(csuppIJ) > 1) csuppJ <- apply(csuppIJ, 2, sum)
if (ncol(csuppIJ) == 1) csuppJ <- sum(csuppIJ)
csuppproj <- t(csuppIJ) %*% rproj
temp <- csuppproj
# Divide rows by mass; and then cols. by sqrt of evals.
csuppproj <- sweep ( sweep(temp,1,csuppJ,FUN="/"),2,
    sqrt(sres$values),FUN="/")

# Value of this function on return: table of projections,
# rows = set of supplementary columns; columns = set of factors.
# (More than 1 supplementary column => labels will be retained.)
# Adjust for trivial factor.
csuppproj[,-1]

}
```

2.7.2 FACOR: Interpretation of Factors and Clusters

Program FACOR, `facor.r`, follows. It will be used in later chapters.

```
facor <- function(kIJ, nclr=nrow(kIJ), nclc=ncol(kIJ), nf=2) {
```

```
# Example of reading input data.
# parm <- read.table("c:/parmenides.dat")
# kIJ <- parm[, -3:-4]

# Definition of total mass, frequencies,
#   profile vectors on I and on J:
k        <- sum(kIJ)
fIJ      <- kIJ/k
fJI      <- t(fIJ)
fI       <- apply(fIJ, 1, sum)
fJ       <- apply(fIJ, 2, sum)
fJsupI   <- sweep(fIJ, 1, fI, FUN="/")
fIsupJ   <- sweep(fIJ, 2, fJ, FUN="/")

# Carry out Correspondence Analysis.
# Diagonalization: definition of factors.
# Following three lines yield matrix to be diagonalized (SVD):
# s_jj' = sum_i (fij * fij') / (fi * sqrt(fj) * sqrt(fj'))
s <- as.matrix(t(fJsupI)) %*% as.matrix(fIJ)
s1 <- sweep(s, 1, sqrt(fJ), FUN="/")
s2 <- sweep(s1, 2, sqrt(fJ), FUN="/")
sres <- eigen(s2)
# Eigenvectors divided rowwise by sqrt(fJ):
evectors <- sweep(sres$vectors, 1, sqrt(fJ), FUN="/")
# Projections on factors of rows and columns
# Note: first column of rproj is trivially 1-valued.
# Note: Observations x factors.
# Read projections with factors 1, 2, ... from cols. 2, 3, ...
rproj <- as.matrix(fJsupI) %*% evectors
temp  <- as.matrix(s2) %*% sres$vectors
# Following divides rowwise by sqrt(fJ) and
#    columnwise by sqrt(eigenvalues):
# Note: first column of cproj is trivially 1-valued.
# Note: Variables x factors.
# Read projections with factors 1, 2, ... from cols; 2, 3, ...
cproj <- sweep ( sweep(temp,1,sqrt(fJ),FUN="/"), 2,
          sqrt(sres$values),FUN="/")

# Create hierarchical clusterings on observations (rows) and variable
# (columns), based on factor projections.  Use weights of obs. or vbe
# We could limit the number of factors used here, e.g. rproj[,2:nlimi
hclr <- hierclust( rproj[,-1], fI)
hclc <- hierclust( cproj[,-1], fJ)
```

```
labsr        <- 1:nclr
labsallr     <- dimnames(fIJ)[[1]]
membersr     <- cutree(hclr, nclr)

labsc        <- 1:nclc
labsallc     <- dimnames(fIJ)[[2]]
membersc     <- cutree(hclc, nclc)

centersr    <- NULL
for (k in 1:nclr) {
  if (length(fI[membersr==k]) > 1)
    centersr <- rbind(centersr, apply(rproj[membersr==k,-1],2,sum))
# Card of cluster = 1 => bypass summing:
  if (length(fI[membersr==k]) == 1)
    centersr <- rbind(centersr, rproj[membersr==k,-1])
  labsr[k] <-  list(labsallr[membersr==k])
}

centersc    <- NULL
for (k in 1:nclc) {
  if (length(fJ[membersc==k]) > 1)
    centersc <- rbind(centersc, apply(cproj[membersc==k,-1],2,sum))
# Card of cluster = 1 => bypass summing:
  if (length(fJ[membersc==k]) == 1)
    centersc <- rbind(centersc, cproj[membersc==k,-1])
  labsc[k] <-  list(labsallc[membersc==k])
}

list(centersr=centersr, cluslabrow=labsr,
     centersc=centersc, cluslabcol=labsc)

}
```

It is handy to have some formatting carried out on the results of FACOR. So we use the program print.facor.r to do this.

```
print.facor <- function(facorres) {

    centr <- facorres$centersr
    labsr <- facorres$cluslabrow
    centc <- facorres$centersc
    labsc <- facorres$cluslabcol

    nclr <- nrow(centr)
    nfac <- ncol(centr)
```

```
nclc <- nrow(centc)
# nfac <- ncol(centc)

cat
("Projections of observations x variables (rows x columns) on \n"
cat(nfac, " factors.  Projections on scale of 0 to 9.\n")

asimple <- round( 9.499*(centr-min(centr))/
                  (max(centr)-min(centr)) )
# asimple <- centr              No scaling
cat
("Obs. (rows) x number of factors retained (on scale of 0 to 9).\
for (i in 1:nclr) {
    cat(asimple[i,], " Cluster ", i, ":", labsr[[i]], "\n")
}

asimple <- round( 9.499*(centc-min(centc))/
                  (max(centc)-min(centc)) )
# asimple <- round( centr )                        No scaling
cat
("Vbes. (cols.) x no. factors retained (on scale of 0 to 9).\n")
for (j in 1:nclc) {
    cat(asimple[j,], " Cluster ", j, ":", labsc[[j]], "\n")
}

}
```

Examples of the use of FACOR will follow in later chapters.

2.7.3   VACOR: Interpretation of Variables and Clusters

Program VACOR, vacor.r, follows. It will be used in later chapters.

```
vacor <- function(kIJ, nclr=nrow(kIJ), nclc=ncol(kIJ)) {

# Example of reading input data.
# parm <- read.table("c:/parmenides.dat")
# kIJ <- parm[, -3:-4]

# Defn. of total mass, frequencies, profile vectors on I and J:
k       <- sum(kIJ)
fIJ     <- kIJ/k
fJI     <- t(fIJ)
fI      <- apply(fIJ, 1, sum)
fJ      <- apply(fIJ, 2, sum)
```

```
fJsupI  <- sweep(fIJ, 1, fI, FUN="/")
fIsupJ  <- sweep(fIJ, 2, fJ, FUN="/")

# Carry out Correspondence Analysis.
# Diagonalization: definition of factors.
# Following three lines yield matrix to be diagonalized (SVD):
# s_jj' = sum_i (fij * fij') / (fi * sqrt(fj) * sqrt(fj'))
s <- as.matrix(t(fJsupI)) %*% as.matrix(fIJ)
s1 <- sweep(s, 1, sqrt(fJ), FUN="/")
s2 <- sweep(s1, 2, sqrt(fJ), FUN="/")
sres <- eigen(s2)
# Eigenvectors divided rowwise by sqrt(fJ):
evectors <- sweep(sres$vectors, 1, sqrt(fJ), FUN="/")
# Projections on factors of rows and columns
# Note: first column of rproj is trivially 1-valued.
# Note: Observations x factors.
# Read projections with factors 1, 2, ... from cols. 2, 3, ...
rproj <- as.matrix(fJsupI) %*% evectors
temp  <- as.matrix(s2) %*% sres$vectors
# Following divides rowwise by sqrt(fJ) and
# columnwise by sqrt(eigenvalues):
# Note: first column of cproj is trivially 1-valued.
# Note: Variables x factors.
# Read projections with factors 1, 2, ... from
# cols; 2, 3, ...
cproj <- sweep ( sweep(temp,1,sqrt(fJ),FUN="/"), 2,
                 sqrt(sres$values),FUN="/")


# Create hierarchical clusterings on observations (rows)
# and variables (columns), based on factor projections.
# Use weights of obs. or vbes.
# We could limit the number of factors used here,
# e.g. rproj[,2:nlimit]
hclr <- hierclust( rproj[,-1], fI)
hclc <- hierclust( cproj[,-1], fJ)


labsr     <- 1:nclr
labsallr  <- dimnames(fIJ)[[1]]
membersr  <- cutree(hclr, nclr)

labsc     <- 1:nclc
labsallc  <- dimnames(fIJ)[[2]]
membersc  <- cutree(hclc, nclc)

centersr  <- NULL
```

```
for (i in 1:nclr) {
  if (length(fI[membersr==i]) > 1)
    centersr <- rbind(centersr,        # Define next row:
      # Marginal aggregate of fIJs, divided by sum of wts.
      apply(fIJ[membersr==i,],2,sum)/
          sum(apply(fIJ[membersr==i,],2,sum)) )
# Card of cluster = 1 => bypass summing:
  if (length(fI[membersr==i]) == 1)
      centersr <- rbind(centersr,
          fIJ[membersr==i,]/fI[membersr==i])
  labsr[i] <-  list(labsallr[membersr==i])
}

centersc  <- NULL
for (j in 1:nclc) {
  if (length(fJ[membersc==j]) > 1) {
    centersc <-
        rbind(centersc, apply(centersr[,membersc==j],1,sum) )
    labsc[j] <-  as.list(j)
  }
  if (length(fJ[membersc==j]) == 1)
            { # Card of cluster = 1 => bypass summing.
    centersc <- rbind(centersc, centersr[,membersc==j])
    labsc[j] <-  list(labsallc[membersc==j])
  }
}


list(vals=t(centersc),labsr=labsr, labsc=labsc)

}
```

The joint interpretation of variables and clusters, VACOR, uses hierarchical clustering. Below we will look at an efficient program for hierarchical clustering, written in C, which can be linked to an R program. Similarly there is a version of program vacor.r which uses this C program. It is called vacor_c.r. We will not provide a listing here because it is similar to vacor.r. Both programs are on the web site.

As for FACOR, it is handy to have a print method for results of VACOR. This we do with print.vacor.r, which will be exemplified in later chapters.

```
print.vacor <- function(vacorres) {

    centc <- vacorres$vals
    labsc <- vacorres$labsc
```

```
labsr <- vacorres$labsr

n <- nrow(centc)
m <- ncol(centc)

cat("Projections of clusters of obs. vs. clusters of vbes. \n")
cat("Projections on scale of 0 to 9.\n")
asimple <- round( 9.499*(centc-min(centc))/
                          (max(centc)-min(centc)) )
# asimple <- round ( 10*centc )          Scale up by 10
# asimple <- centc                       No scaling

cat("Var. labs.: ")
for (j in 1:m) cat(labsc[[j]], " ")
cat("\n")
for (i in 1:n) {
    # cat(as.matrix(asimple[i,]),
    #     " Cluster ", i, ": ", labsr[[i]], "\n")
    # cat(i, "&", labsr[[i]], " & ", asimple[i,1],
    #     " & ", asimple[i,2],
    #     " \\\\ \n")
    cat(i, " & ")
    for (j in 1:m) {
        cat(asimple[i,j], " & ")
    }
    cat("\\\\ \n")
}

}
```

2.7.4   Hierarchical Clustering in C, Callable from R

As mentioned in Chapter 1, for any way sizable n, number of rows, an environment like R will have problems with iterative algorithms, due to the priority given by this environment to program interpretation as opposed to compilation. A way to handle a large n situation is to use a C program and link this into R. Such a program is run in the following way.

```
HierClust <- function(arr, wts)
{
# Example of use.
# dyn.load("C:/hierclust.dll")
# arr <- matrix(c(1,4,2,3,0,2,1,2,4,2,3,2),nrow=4,ncol=3)
# wts <- as.single(c(0.7, 0.3, 0.5, 0.8))
```

```
# NOTE: Be careful with input data.  read.table produces an object of
# type list.  Convert this to a matrix (e.g.  y <- as.matrix(x) ).

order <- as.integer(rep(0,nrow(arr)))
ia <- as.integer(rep(0,nrow(arr)-1))
ib <- as.integer(rep(0,nrow(arr)-1))
height <- as.single(rep(0,nrow(arr)-1))

output <-   .C("HierClust", as.integer(order),
                            as.integer(ia),
                            as.integer(ib),
                            as.single(height),
                            as.single(arr),
                            as.single(wts),
                            as.integer(nrow(arr)),
                            as.integer(ncol(arr)))


retlist <- list(merge=cbind(output[[2]], output[[3]]),
                height=output[[4]],order=output[[1]])

retlist

}
```

We will assume that we have the driver script, shown in full above, in file
`hcluswtd_c.r`. We will further assume that the C program is in `hierclustx.c`,
and that on Windows systems a linkable object program called `hierclustx.dll`
has been created from this. On Unix systems, an object file, `hierclustx.o`,
is used.

Reading in and using these programs is carried out as follows.

```
# Load this, in Intel/Windows:
dyn.load("c:/hierclustx.dll")
# Ignore the warning message ("DLL attempted to change...").
# Source the R driver:
source("c:/hcluswtd_c.r")
# Use Fisher's iris data, with identical weights for each iris flower.
data(iris)
x <- iris[,1:4]
# Here we have to coerce to matrix data type.
xh3 <- HierClust(as.matrix(x), rep(1/150, 150) )
```

In the S-Plus dialect of R (S-Plus is the commercial product, for which free
student versions are available), on a Unix/Solaris machine, we do the same
operations. First to compile (but not link): `gcc -c hierclustx.c`. Then:

```
# In S-Plus:
source("hcluswtd_c.r")
dyn.open("hierclustx.o")
x <- rbind( iris[,,1], iris[,,2], iris[,,3] )
xh4 <- HierClust(x, rep(1/150, 150))
```

On both R on a Windows platform (Intel assumed), and S-Plus on a Sun Sparc Solaris platform, we plotted the output of the callable C code with `plclust`.

2.8 Summary

Correspondence analysis displays observation profiles in a low-dimensional factorial space. Profiles are points endowed with χ^2 distance. Under appropriate circumstances, the χ^2 distance reduces to a Euclidean distance. A factorial space is nearly always Euclidean.

Simultaneously a hierarchical clustering is built using the observation profiles. Usually one or a small number of partitions are derived from the hierarchical clustering. A hierarchical clustering defines an ultrametric distance. Input for the hierarchical clustering is usually factor projections.

In summary, correspondence analysis involves mapping a χ^2 distance into a particular Euclidean distance; and mapping this Euclidean distance into an ultrametric distance. The aim is to have different but complementary analytic tools to facilitate interpretation of our data.

3

Input Data Coding

3.1 Introduction

Issues relating to coding of data, prior to data analysis, have been addressed far less by statisticians and data analysts than the methods themselves. Input data coding belongs to the undergrowth, so to speak, of data analysis. However input data coding is a highly practical aspect of the analysis of data since it impacts directly on the result obtained. It is therefore instructive to look at innovative and perhaps even unusual techniques in regard to data coding.

We will overview various interesting practical techniques. Input data coding has always been accorded considerable importance in the framework of correspondence analysis. We consider these lessons and their implications to be valuable. This work is also potentially of interest to other domains of related multivariate data analysis methods – data mining and machine learning, neural networks and pattern recognition, to name but a few.

One tradition in data analysis links the analysis method with input data "levels of measurement." This is the typology of measurement scales introduced by S.S. Stevens in the 1940s for use in psychophysics, and based on a measurement value being of nominal, ordinal, interval or ratio scale. The appropriate analysis methods depended on the input data type, the level of measurement. If metric assumptions regarding input data were not warranted, then this had implications for implanting the data in a metric space.

Stevens's categorization of data has been influential. It has been used as a set of guidelines on appropriate methods to use in specified circumstances. Velleman and Wilkinson [84] criticized it on such grounds as the following: imprecision in practice; insufficiency for describing multidimensional characteristics of scales; and the fact that measurement scales can be a matter of interpretation. They found Steven's *prescriptions*, and the guidelines associated with his data measurement taxonomy, often to be unjustified *proscriptions*.

If we had to characterize in one property how correspondence analysis differs from other similar analysis approaches then it would be its flexibility in handling input data. However the input data coding issue is very important in correspondence analysis. By coding the data in a particular way, the χ^2 distance on profiles becomes the classical Euclidean distance. This example shows how coding has immediate and direct implications for the analysis we are carrying out.

Input data coding is commonplace with data analysis algorithms. Consider the need for data normalization or standardization, to prevent some variables or observations from "shouting louder than others." Data scaling or normalization are commonly used prior to analysis of real-valued (continuous, quantitative) data. There are various normalization and rescaling schemes. Milligan and Cooper [58] review a number of these and, for cluster analysis, they present simulation results which favor standardization through subtracting the minimum value, and dividing by the range. Correspondence analysis ties together closely such normalization and the nitty-gritty of the analysis algorithm. An observation's or variable's mass, for example, is usually determined from the given data (as a marginal). Therefore an understanding of the correspondence analysis algorithm is needed in order to appreciate the implications for input data coding.

3.1.1 The Fundamental Role of Coding

The theme of a study sometimes requires analyzing a contingency table k_{IJ} where neither the sets I and J nor the numbers $k(i, j)$ should be other than they are. Usually however in collecting data the construction of the tables to be analyzed, and the spatial representations which one associates with the data, lead to various possibilities among which the data analyst has to choose. The properties of homogeneity, exhaustivity, fidelity of geometric representation and stability of results are discussed in Bastin et al. [12].

Homogeneity: The theme of the study delimits the domain from which one collects the data. The point of view of the study fixes the form of this data. That is to say, it fixes the level at which one describes reality: spatial dimensions, chemical composition, word counts, etc. However in practice it is often necessary to analyze heterogeneous sets of variables collected on different levels: qualities, integers; continuous quantities of different natures or orders of magnitudes. One tries in such cases to arrive at some measure of homogeneity using mathematical transformation or coding. By taking each variable as a question containing a finite set of response modes (which is strictly the case for a qualitative variable; and will be also for a quantitative variable if the interval of variation is partitioned into classes) we end up with a quasi-universal coding format: the questionnaire. It is also possible to balance the role of different groups of numerical variables using weighting.

Exhaustivity: To determine through analysis how a certain level of reality is ordered, in accordance with which axes it is necessary to have taken this level in its totality, or at least to have extracted a sample of uniform density. From this point of view, Louis Guttman considered every finite questionnaire as an extraction from a continuous universe of possible questions: hence the importance of continuous models. In the analysis of such a model, an essentially one-dimensional situation can produce an infinite sequence of factors.

We approximate an exhaustive description by a nomenclature which is more and more fine-grained. We can regard such a nomenclature as a partition of a space into a mosaic of non-overlapping classes (e.g. the set of all human activities is broken up into socio-professional categories). But in practice the partition is approximate, and the classes are fuzzy. The principle of distributional equivalence guarantees that cumulative rows or columns of neighboring profiles in a table changes the results very little. What is more, if starting with a cloud $N(I)$ we form aggregated rows or columns arbitrarily (which therefore could include distant rows or columns), the cloud of centers of these aggregates has the same principal axes of inertia as $N(I)$, but with weaker moments of inertia.

Fidelity of the geometric representation: Algorithmic calculations yield tables of values and planar maps on the basis of which we recognize, as far as this is possible, the structure of a multidimensional object $N(I)$. In addition, this object has to be a faithful geometric representation of the system of properties and of the observed relations. If the coding imposes the structure of an a priori hypothesis on the primary data then this hypothesis will be found from the analysis as an artifact. To address this issue, it is generally good to base the analysis on symmetric roles for a quality and for its opposite.

Universality of processing: By coding all data according to the same format, i.e., a correspondence table, one can in very different domains apply the same analysis algorithms; and following the analysis one can critique on the basis of many cases the conclusions suggested by the tables and graphics. From this point of view, coding of preference data in the form of a correspondence table is interesting.

Stability of results: As we have said, in the same study different approaches seem to be possible. It is particularly satisfactory if all approaches point in the same direction, and give similar results. For this reason, it may be useful to consider a number of codings of the same set I, for a given cloud $N(I)$. In practice, as opposed to sample problems, it is a matter of approximate equality of results. In particular this leads to reduced coding which simplifies the data to be analyzed without changing greatly the outcome.

3.1.2   "Semantic Embedding"

Correspondence analysis is based on a range of input data coding options. The following view of data coding is outlined in Benzécri [29].

As far as the conceptual understanding of numbers is concerned, the distinction between qualitative and quantitative appears to us to be often misunderstood. In brief, one should not say: a continuous numerical value equates approximately a quantitative datum, and a value taking a finite number of modalities equates approximately a qualitative datum. Such assertions are not true because at the level of the statistical unit (e.g., a patient record) a numerical datum such as age or even pulse rate or glycemy is not generally to be taken with given precision, but rather in accordance with its significance. From this point of view, there is no difference between age and profession.

To compare one observation to another, one should not consider two sets of primary data, for example two sets of 100 real numbers, or a point in $\mathbb{R}^{100}$ compared with another point in $\mathbb{R}^{100}$ between which there are no global or overall resemblances to be seen. Instead one should consider the synthesis of these sets, leading to some gradations of interpretation, or to discontinuities, and ultimately to diagnoses.

As far as calculation is concerned, correspondence factor analysis algorithms – implying, as is well-known, the numerical cost of matrix diagonalization – and hierarchical clustering, operate on the coded data following the global principle of distributional equivalence of profiles. This scores much in efficacy compared to the processing of contiguities between observations, used by stochastic approximation algorithms. The latter, nowadays, are often in the guise of neural networks.

To summarize: coding such as complete disjunctive coding maps the initial data value into new data values which have greater semantic content. Secondly, the distributional equivalence principle, and the simultaneous accounting for variation and relationship in observations and attributes, are directed towards the incorporation of semantic information.

Motivation for the type of "semantic lifting" to be seen in various forms of input data coding below include the following:

- We seek to give the observations constant mass, thereby normalizing the input data.

- The space of observation points is sometimes boolean valued, i.e., the cloud's points are placed at the vertices of a hypercube. In any case, there is greater isotropy in the cloud compared to the initially given values.

- Notwithstanding the previous points, we find that a low dimensional projection, before and after, gives a relatively good match. In other words, the recoded data is "similar to the original data."

Correspondence analysis facilitates use of very different types of observation, and very different types of attribute, through the supplementary element mechanism. This can be done in other ways, of course, such as through canonical correlation analysis, or CCA. CCA involves finding the principal axes in two spaces, using the same observation set, with the constraint of orthogonality link between the two sets of principal axes. Supplementary elements are on many levels more straightforward, and more integral to the analysis.

Let us look at one example of the need to analyze attributes of different natures. The efficient market hypothesis states that a price y_t at time point t is a martingale, i.e., $E\{y_{t+1} \mid y_0, y_1, \ldots, y_t\} = y_t$, where E is expectation. This appears to exclude on theoretical grounds any possibility of forecasting the price movement. However it may well be possible to predict over long time horizons by using the information present in additional time series such as earnings/price ratios, dividend yields and term-structure variables [56].

Could we approach the problem of predicting financial data, and bypassing the efficient market hypothesis, by coding alone? Ross [75] avails of exactly such a stratagem. Using differenced lognormal price values, he defines states – called modalities below – with the following properties. Firstly, the data range is roughly −0.06 to +0.06. States are defined for values less than −0.01, between −0.01 and 0, between 0 and +0.01 and greater than +0.01. It is shown, then, that there is structure to be found in the data when coded in this way.

In section 3.2.5 we use data coding to find patterns in which environmental data initially appears to contain no interesting patterns.

3.1.3 Input Data Encodings

A range of practical traditions have been built up around correspondence analysis. First some comments on correspondence analysis are in order. Why *might* lessons learned from it be generally valid when using other analysis methods?

Coding of input data is especially important in the case of correspondence analysis, since it is best seen as a privileged method for the analysis of data types other than quantitative: qualitative or categorical, logical or binary, frequency data, mixed quantitative/qualitative, etc. Correspondence analysis differs from principal components analysis in that the χ^2 distance replaces the Euclidean one. Viewed from a geometric and data exploration perspective, background texts include [16, 23, 24, 25, 44, 50, 68, 76]. Just a few further references on particular topics in this area include: [17, 45, 49, 51, 71, 85].

Correspondence analysis, it could be claimed, is particularly appropriate for analyzing contingency tables, but has also found wide application to many other types of data. In most such cases, some kind of transformation of the data is required depending on the nature of the data. We will consider some of these techniques and practices in regard to data transformation, which we call *data coding*.

We will use the terminology of correspondence analysis. Terms which we will discuss include: doubling; the lever principle; complete disjunctive form; fuzzy and piecewise linear coding; and the personal equation. Some terms used are well covered in basic texts on correspondence analysis, and in chapter 2, and we will not seek to define or illustrate them. Such terms include: mass (i.e., weight); barycenter (center of gravity); Burt table (the matrix product of the transpose of a contingency table with itself); and supplementary rows or columns (data adjoining the input data, and projected onto the output planar representations following the correspondence analysis).

TABLE 3.1

Scores (out of 100) of 5 students, A–E, in 6 subjects.

	CSc	CPg	CGr	CNw	DbM	SwE
A	54	55	31	36	46	40
B	35	56	20	20	49	45
C	47	73	39	30	48	57
D	54	72	33	42	57	21
E	18	24	11	14	19	7
	CSc	CPg	CGr	CNw	DbM	SwE
mean profile:	.18	.24	.12	.12	.19	.15
profile of D:	.19	.26	.12	.15	.20	.08
profile of E:	.19	.26	.12	.15	.20	.08

3.1.4 Input Data Analyzed Without Transformation

As a basis for the discussion to follow on data coding, we will summarize the types of input data which we are given to analyze.

Contingency tables: Frequency tables (arrays) and cross-classifications come under this heading. In a contingency table, the total of a row or of a column is usually quite meaningful. In such a table, we can also calculate the relative frequencies by dividing the number in a cell by the total for the table. Contingency tables are analyzed as such by correspondence analysis.

Description tables: A set of individuals or objects I described by a set J of variables gives rise to a table of descriptions (real values or integer values). When J is a homogeneous set of measurements we can apply correspondence analysis directly. A remark though is in order: doubling and other forms of coding – to be discussed below – can allow the χ^2 metric used to be transformed to a weighted Euclidean metric. In practice, it may be advisable to carefully check out the various alternative possibilities available to the analyst in doing this.

Mixed qualitative and quantitative data: Qualitative data are also referred to as categorical (including binary) or symbolic data. What was indicated under description tables above applies also here.

Tables of scores: In market research surveys, respondents may be asked to attribute scores on a rating scale, say from 0 to 10, to the different qualities of a detergent, and the results are cross-tabulated giving rise to a table of scores. We might consider such a table as a contingency table, for the sum of a row and the sum of a column have a meaning, e.g., in a students × subjects array, the sum of the scores awarded to a student, or in a subject. However in such a table, certain examiners consistently award higher scores, leading to a higher mass being associated with these exams. Conversely the examiners who give smaller scores will have a low mass. It is therefore preferable to use a procedure which gives to each examiner the same mass.

Based on any of the foregoing types of data, correspondence analysis high-

TABLE 3.2
Doubled table of scores derived from Table 3.1.

	CSc+	CSc-	CPg+	CPg-	CGr+	CGr-	CNw+	CNw-	DbM+	DbM-	SwE+	SwE-
A	54	46	55	45	31	69	36	64	46	54	40	60
B	35	65	56	44	20	80	20	80	49	51	45	55
C	47	53	73	27	39	61	30	70	48	52	57	43
D	54	46	72	28	33	67	42	58	57	43	21	79
E	18	82	24	76	11	89	14	86	19	81	7	93

lights the similarities and the differences in the profiles. Take for example the table of scores shown in Table 3.1. Subjects are: CSc: Computer Science Proficiency, CPg: Computer Programming, CGr: Computer Graphics, CNw: Computer Networks, DbM: Database Management, SwE: Software Engineering.

Notice that (i) all the scores of D and E are in the same proportion (E's scores are one-third those of D), and (ii) E has the lowest scores both in absolute and relative terms in all the subjects. Without data coding, correspondence analysis of this table would show, for example, D and E at the same location in the map because they have identical profiles.

Both show a positive association with CNw (computer networks) and a negative association with SwE (software engineering) because in comparison with the mean profile, D and E have, in their profile, a relatively larger component of CNw and a relatively smaller component of SwE.

Is this what we hoped to discover from a correspondence analysis? What, on the other hand, would be of interest to us is the real differences between the students and how these differences arise. We therefore need to clearly differentiate between the profiles of D and E, which we do by *doubling* the data.

3.2 From Doubling to Fuzzy Coding and Beyond

3.2.1 Doubling

We use a procedure called doubling (*dédoublement*): we attribute two scores per subject instead of a single score. The "score awarded," $k(i, j^+)$, is equal to the initial score. The "score not awarded," $k(i, j^-)$, is equal to its complement, i.e., $100 - k(i, j^+)$, so that

$$k(i, j^+) + k(i, j^-) = 100$$

The table is then doubled, by associating with each subject j the two columns j^+ and j^-. All the rows now have the same total, hence each student receives the same mass (see Table 3.2).

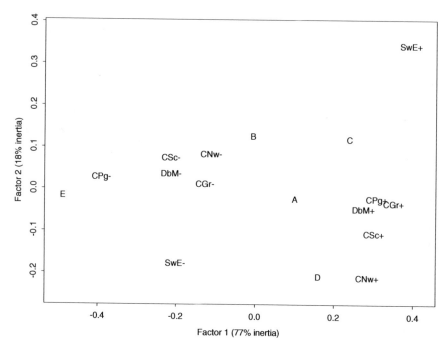

FIGURE 3.1

First two factors of correspondence analysis of data in Table 3.2.

In the correspondence analysis display of this table (see Figure 3.1), we obtain one point for each student and two points for each subject, one representing the favorable score attributed (j^+), and the other the unfavorable score (j^-). The points $k(i, j^+)$ and $k(i, j^-)$ have their barycenter at the origin of the display. We can now see the clear difference between E and all the rest on the one hand, and the association of E with all the negative axis scores on the other.

The *lever principle* is used in the interpretation of the low-dimensional projection of a doubled table in correspondence analysis.

A "+" variable and its corresponding "−" variable lie on the opposite sides of the origin and collinear with it. If the mass of the profile of j^+ is greater than the mass of the profile of j^- (which means that the average score for the subject j was greater than 50 out of 100), the point j^+ is closer to the origin than j^-. This is called the lever principle. We can thus read from the map that except in CPg, the average score of the students was below 50 in all the subjects.

To summarize, whenever we are dealing with a table of scores, it is important to remember that correspondence analysis compares profiles. This will automatically lead us to consider adopting the doubling principle in dealing with such data.

Cumulating Complementary Columns

Sometimes it is more expedient to replace, in a doubled table of scores, all the columns relating to the "−" variables by a single column obtained by cumulating them, thus obtaining a single point "nega" in the graphs in place of all the "−" variables [1, 41]. Individuals who are generally strict in attributing scores will then be placed relatively closer to the point "nega." In [41] analysis of the personal particulars of the individuals suggested that housewives with several children tended to be liberal in attributing scores while housewives with no children tended to be stricter (i.e., closer to "nega").

Doubling and complete disjunctive form (to be discussed below) offer a solution to another problem, that of variable weighting. We see that various types of coding described, including doubling, complete disjunctive form, and fuzzy coding, all lead to equally-weighted observations. At the expense of more storage space, we see therefore that the data become considerably more well-behaved and interpretable.

3.2.2 Complete Disjunctive Form

The responses of a set of subjects to a set of questions are coded as boolean (or logical) values, and this format also serves to code several other types of data, including complete disjunctive form, which is defined by a set of response variables which sum to a constant.

Let I be a set of subjects i, Q a set of questions q and J_q a set of the response categories corresponding to the question q. We suppose that the response of any subject to a question q falls under one of the categories J_q. J is the union of all the J_q, for q belonging to Q, i.e., J is the set of all the response categories pertaining to all the questions.

k_{IJ} is the table of responses. With each individual, a row of the data table is associated. To each question q there corresponds a block J_q of columns. $k(i,j) = 1$ if the subject i chooses the category j, and zero otherwise. Hence in the row i in each block J_q there is a 1 in the column pertaining to the response category j chosen by the subject for the question q, and zeros elsewhere. The total of each row of the table k is therefore equal to the number of questions.

Description tables in which a set I of individuals is described by discrete qualitative variables (e.g., sex: 2 categories) are coded in this format. More generally, every quantitative variable can be transformed into a qualitative variable if the interval of variation is divided into class intervals and coded in the complete disjunctive format. Sometimes the class intervals are of equal width, sometimes they are so chosen that the resulting categories have approximately equal numbers of observations in each of them, and sometimes the limits are chosen in order that the resulting categories have a well-accepted significance, as in clinical data, and so on. The categorization of a continuous quantitative variable into class intervals is not always easy (cf. below: piecewise linear coding).

When the data consists of a mixture of quantitative and qualitative variables the quantitative variables can be transformed into qualitative variables and the resulting table can then be put into the (0,1) format.

Remark: Analysis of Tables with 0 and 1 Values

The analysis of a table $I \times J$ in complete disjunctive format furnishes, for the set of categories J, principal coordinates which (within a constant coefficient) are the same as those obtained by analyzing the Burt table k'_{JJ} derived from $I \times J$ in the following manner.

Let $k'(j,j') =$ the number of individuals i of I belonging simultaneously to both the categories j and j'. The Burt table is a true contingency table. Therefore we have a clear link between correspondence analysis of such a contingency table and correspondence analysis of a table with 0 and 1 values.

The Burt table can be analyzed as the principal table, and we can adjoin to it the table in $I \times J$ as supplementary. In this way we get the principal coordinates of both J and I and the simultaneous representation of both sets in the output representations.

3.2.3 Fuzzy, Piecewise Linear or Barycentric Coding

For continuous quantitative variables, a fuzzy coding scheme is often preferred to complete disjunctive coding for the following reasons. Complete disjunctive

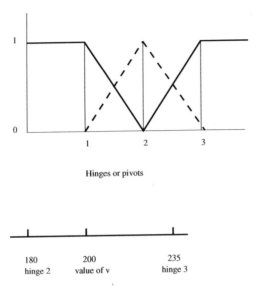

1   2   3

Hinges or pivots

180	200	235
hinge 2	value of v	hinge 3

FIGURE 3.2
Hinges used in piecewise linear (or fuzzy, or barycentric) coding. We have hinges $v_2 = 180$ and $v_3 = 235$.

coding gives rise to a sharp discontinuity. Instead progressive passage from one category to another requires that values close to the border separating two categories are not remarkably different in their recoded values. This leads to fuzzy coding schemes (Gallego, [43]). The main difference noticed by Gallego between disjunctive and fuzzy coding is the clarity and facility of interpretation of the graphical representations associated with fuzzy coding, the regularization of the paths associated with the variables resulting from the reduction of the distance between contiguous categories, leading to a more clear highlighting of the relationship between the categories of the various variables.

One scheme of fuzzy coding found useful by Gallego was "semilinear" coding, and is fully described in [21] under the name *piecewise linear coding* (also *barycentric coding*). Here piecewise means "in each interval" (i.e., between two consecutive hinges, or reference values, cf. Figure 3.2).

The principle is to choose, for each variable, a set of 3 or 4 reference values called hinges. An equal number of categories are constituted around these hinges. Any value of a variable less than the lowest hinge will be coded as 1 in the lowest category, and as 0 in all the other categories. Any value of the variable which is greater than the highest hinge will likewise be coded as 1 in the highest category, and as 0 in all the other categories. Any value falling between two consecutive hinges will be distributed between the two corresponding categories and coded as discussed next.

Formula for Barycentric Coding using Hinges

Let, for example, the hinges be (125, 180, 235) for a variable which we shall refer to as v (see Figure 3.2). We shall name the three categories corresponding to the three hinges as v_1, v_2, v_3. How will the value 200 be coded? The value 200 lies between the second and the third hinges, therefore the first category, v_1, is zero, and the second and third are non-zero.

The value 200, lying between the middle and last hinges, can be considered as the barycenter (weighted average) of these two hinges with appropriate masses adding up to 1. The value 200 is at 20/55 units from the second hinge 180, and 35/55 units from the third hinge 235. Hence the category v_2 and the category v_3 will receive respectively $35/55 = 0.64$ and $20/55 = 0.36$. The value 200 is therefore coded as (0, .64, .36).

In this scheme, there are as many categories as there are hinges; and whatever the number of hinges, there are at most only 2 categories different from 0. The values of the categories are either positive or 0, and their total is 1. This barycentric coding may be preferred to the (0,1) coding when observations are few in number, and categorizing them with a view to complete disjunctive coding will result in one or several of them having too few observations.

To choose the hinges, one approach is to look at the histograms of a variable. The histogram modes are noted, as are outlying values, in choosing the locations of hinges or pivots. The case of more than 3 hinges can be considered on similar lines.

While some precision is lost in this coding scheme, it has the advantage of being clear and simple. Secondly, it is important for the stability of the analysis that there should not be extreme categories of very low mass, and hence sensitive to the fluctuations of the sampling.

An alternative automatic choice of hinges is used in extended examples below, based on quantiles. This ensures robustness, and generality for all histograms or distributions.

An Example of Fuzzy Coding

An example of piecewise linear coding is given in [21]. Eight new high performance plastics and their fiber glass reinforced versions, making a total of 16 thermoplastics, are tested for five properties. See Table 3.3. The data are not enough to describe the materials fully, but it is precisely the small quantity of the data that has dictated the necessity of using barycentric coding. The authors [21] also say that various trials not reported in the article led to the conclusion that this coding gives the results a robustness that is not sensitive to the imprecision of other coding schemes.

The first two variables are temperatures and the last three are resistances measured in various units. Hence the variables are quantitative and heterogeneous. The quality of the product is judged by higher values of the variables.

TABLE 3.3

Example: superplastics data.

	Tflx	Tvll	Rtrc	Rflx	Rchc
ps	174	150	65	27	70
PS	180	170	108	75	85
pei	200	170	105	33	50
PEI	210	170	160	83	100
pes	200	180	90	30	84
PES	230	200	140	85	80
pcl	170	160	230	90	270
PCL	230	220	170	160	130
hta	230	200	85	25	123
HTA	252	200	135	85	59
PPS	270	210	160	145	80
pai	278	220	195	46	134
pek	150	250	100	40	80
PEK	315	250	160	100	95
pa6	80	100	90	25	40
PA6	255	130	170	85	120

The hinges are at first tentatively chosen by arranging the values of the variable in increasing order, and specifying the hinges by their ranks:

```
Tflx: ranks (2,8,15), i.e. the values (150,210,278)
Tvll: ranks (2,8,15), i.e. the values (130,180,250)
Rtrc: ranks (2,9,15), i.e. the values (85,135,195)
Rflx: ranks (4,8,15), i.e. the values (30,75,145)
Rchc: ranks (3,8,15), i.e. the values (59,84,134)
```

However this coding is not considered to reflect some of the extreme values faithfully since, for example it transforms both the extreme values of Tfl, viz. 80 for pa6 and 150 for pek into (1,0,0). Hence the hinges need to be modified to achieve a more satisfactory coding. For this, several histograms for the variable Tfl are inspected in order to see how far away the extreme values are situated from the rest of the distribution, which in turn helps determine where to locate the hinges.

The modified hinges for Tflx are (120, 210, 300), using which the values 80 and 150 are respectively coded as (1, 0, 0) and (0.66, 0.33, 0), instead of both being coded as (1, 0, 0). This is a satisfactory improvement on the first trial.

```
Tflx: modified hinges (120, 210, 300)
```

In the same way hinges are modified for the other variables also.

```
Tvll: modified hinges (125, 180, 235)
Rtrc: modified hinges (80, 140, 200)
```

TABLE 3.4

Barycentric recoding of ordered
boolean data.

category	c1	c2	c3	c4	c5
i1	1	0	0	0	0
i2	0	1	0	0	0
i3	0	0	1	0	0
i4	0	0	0	1	0
i5	0	0	0	0	1

cat1	cat2	cat3
1	0	0
1/2	1/2	0
0	1	0
0	1/2	1/2
0	0	1

```
Rflx: modified hinges (28, 78, 128)
Rchc: modified hinges (55, 85, 150)
```

While the middle hinge in most cases falls at or near the median value, it is the extreme hinges which have to be determined with care as all the values beyond these hinges are not distinguished.

The correspondence analysis results can be found in [20]. Further examples of piecewise linear coding can be found in [32].

Barycentric Recoding of Data Already Boolean Coded

Frequently it is useful for interpretational convenience to recode into three categories cat1, cat2, cat3 a variable already coded into (0,1) across ordered categories, e.g., 5 categories c1, c2, c3, c4, c5. The schema shown in Table 3.4 comprising 5 individuals i1 to i5, belonging each to a single category cx, will be used for illustration.

This recoding is achieved if we first convert the boolean table into a table M of category numbers, and then apply to M a piecewise linear coding using 1,3,5 as the values of the hinges.

One interesting conclusion is that fuzzy or piecewise linear coding can be used for (i) robustifying the analysis, or (ii) for modifying logical coding to have no sharp transition. Subsequently, then, other considerations like roughly equal masses of the categories come into play.

3.2.4   General Discussion of Data Coding

A long tradition of input data coding is associated with correspondence analysis [16, 17, 23, 24, 25, 44, 45, 50], and the journal *Les Cahiers de l'Analyse des Données*. One reason why coding is so important is related to the observation weighting scheme used in correspondence analysis. In principal components analysis (formally, a very similar data analysis method based on eigen-reduction to determine principal axes of inertia in the multivariate space of observations endowed with a Euclidean metric) the weights of observations are often taken to be equal, and the weighting of variables is carried out through standardization. The latter involves centering to zero mean, and reducing to unit standard deviation, which has as a consequence that proximity relationships between variables are, in fact, correlations.

In correspondence analysis there is no such standardization, nor weighting. But an analogous effect is obtained by an inherent scheme, based on masses of both observations (rows) and variables (columns). A set of observation or row masses is defined as the averaged column totals. Similarly a set of variable or column masses is defined as the averaged row totals.

Consider an example of a correspondence analysis of examination results, with the total mark for examination 1 as 100, the total mark for examination 2 as 150, and the total mark for examination 3 as 100. A student's score was, we will say, $(71, 98, 65)$. Now consider an expanded student (observation) vector of scores defined as $(71, 100 - 71, 98, 150 - 98, 65, 100 - 65) = (71, 29, 98, 52, 65, 35)$. Note how doing this will necessarily provide each student with the same mass, which is based on the sum of these values. Through the use of additional or augmented examination scores or marks, this scheme provides identical weighting for all students.

Interpretation of results must then face the issue of twice as many variables. For one examination, let's say on databases, there is a variable called "Databases" and another one called "Not Databases." The former represents how well the student has done, while the latter represents the student's lack of knowledge. Sometimes this can be useful for interpretation. Consider for instance a case where we find projections on the first factor to be high for mathematical subjects and also to be high for "Not" linguistic subjects.

The coding used in this example is referred to as doubling. In output (low-dimensional space) projections, a so-called law of the lever principle can be seen for pairs of such doubled variables.

A form of coding which takes this approach further is known as complete disjunctive form coding. Singleton fuzzification of variable $x = 3$, writing it as a unary relation m on the set $X = \{1, 2, \ldots 9\}$, is:

$$m(x) = 0\ 0\ 1\ 0\ 0\ 0\ 0\ 0\ 0$$

In complete disjunctive form coding it is assumed that just one category of this variable (category 3 above) will have a 1 value. It follows that this recoded variable will contribute a constant to the row total. So again we achieve exact equality of row masses. Another use of such coding is to accumulate together (simply by summing) some of the "Not" variables into one overall "Not" variable. This could be appropriate when the "Not" variables refer to missing values – we can avoid taking missing values into account for each variable and instead take an overall missing value into account. Studies using accumulated variables can be found in [1, 41].

The cross-product array, crossing variables × variables, is known as a Burt table. Analysis of data coded in complete disjunctive form is known as multiple correspondence analysis.

The coding introduced thus far provides examples of fuzzy singleton coding. In previous sections we have noted the need for this type of coding, but also that fuzzy singleton coding is not precise enough in its own right. Fully fuzzy coding has been introduced in correspondence analysis under the names of fuzzy, piecewise linear or barycentric coding [21, 43].

The main difference noted by Gallego [43] between disjunctive and fuzzy coding is the clarity and facility of interpretation of the graphical representations associated with fuzzy coding, leading to a more clear highlighting of the relationship between the categories of the various variables.

The singletons used to define a fuzzy partition have been referred to, in the correspondence analysis literature, as hinges or reference values. In correspondence analysis, just as in mainstream fuzzy control and logic, the triangular-trapezoidal form of fuzzy subsets has proved popular. For continuous support intervals, again implementation of fuzzification is very straightforward.

Interpretation of the analysis results relies on defuzzification, or even on not fuzzifying the output at all. Fuzzy correspondence analysis fits into this latter category. It can be characterized as the mapping of fuzzy sets to crisp sets. The motivation for fuzzification is that it is desirable – maybe even necessary – in practical problem solving, to facilitate data interpretation.

3.2.5   From Fuzzy Coding to Possibility Theory

The choice of hinges in our work in [64] was made on the basis of other considerations. Rather than seeking to robustify the analysis, by taking the extreme hinges somewhere around the 5th and 95th percentile values, the interpretability of these extreme values was of interest, and was preserved.

The study related to the quality of astronomical observations. So-called *seeing* is the observing quality and is usually measured with a Gaussian fit to the blur of a star on the detector. The blur is caused by the earth's atmosphere effecting the path of the light coming from the distant cosmic object. Seeing is taken as a single value, the "full width at half maximum" of the approximately Gaussian shape characterizing the blur. For a 2-dimensional strictly Gaussian distribution, the full width at half maximum is a constant times the standard deviation.

The dependence of seeing on a number of meteorological and environmental variables was investigated. The variables used are listed in Table 3.5. Nightly

TABLE 3.5

Fuzzy variables used in *seeing* study.

s1H	windspeed (ms^{-1}) – High
s1L	windspeed – Low
s1dH	windspeed standard deviation – High
s1dL	windspeed standard deviation – Low
rhH	relative humidity (%) – High
rhL	relative humidity – Low
t1H	temperature (^0C) – High
t1L	temperature – Low
pH	pressure (mB) – High
pL	pressure – Low
seeH	seeing (arcsec) – High
seeL	seeing – Low

median values were used, on the basis of data collected at observing sites in Chile from 1989 to 1992. Following rejection of records with missing values, around 600 records remained. Correcting for seasonal differences is important: for this reason, the correspondence analyses were carried out on the seasons separately. This led to the analysis of 11 seasons' data, with around 60 records analyzed in each season.

For seeing, and for most of the other variables, "high" and "low" are intuitively clear. Rather than robustifying by mapping a few extreme values to the same value, we sought to transform the data so that easily interpretable extreme values were focused on. This of course leads also to a robust analysis, since extreme values are not allowed to unduly influence the analysis.

Another reason can be adduced for the use of values "low" (1,0), "high" (0,1) and "moderate" (x, y) with $x = 1 - y; x, y > 0$. This lay in our hope to pursue the analysis in order to forecast astronomical seeing. We hoped to proceed in the following stages: (1) ascertain if seeing had a clear dependence on the other meteorological variables; and, if so, determine this dependence. Then, (2), obtain predictions of the meteorological variables (using meteorological modeling techniques), and obtain the corresponding seeing value. To obtain predictions of meteorological variables is easier if all we have to do is obtain predictions of "low," "moderate" or "high" values. The role of correspondence analysis in the first of the stages referred to above was to investigate the roles played by the variables, and to plan the subsequent stages of the analysis (involving, e.g., linear or nonlinear regression). Thus, correspondence analysis was used as a method of exploratory data analysis, by means of which we sought to unravel the complexities or the interesting aspects of the data.

Hinges were defined at the 33rd and 67th percentile values of each variable. Piecewise linear interpolation was used for intermediate ("moderate") values.

Examples of the coding follow. Consider a value of pressure of 744.0 mB.

For a given season's values, the 33rd and 67th percentiles are found to be 742.4 and 743.8. Then our value of 744.0 is coded (0, 1) (0 times "low," 1 times "high"). Consider next a value of 741.6. It is coded (1, 0) (1 times "low," 0 times "high"). Finally consider an intermediate value, 743.1. It lies exactly half way between the thresholds for "low" and "high," and so this value is coded (0.5, 0.5).

As mentioned above, correspondence analyses were carried out on 11 seasons, from spring 1989 to spring 1992, and are shown in Figures 3.3 and 3.4. To be consistent with the fuzzy coding used, we must generalize empirical probabilities to possibilities. Two measures are of greatest interest: the conditional possibility of good (i.e., "low") seeing; and the joint possibility of the variables happening. The first of these is an expression of relatedness; the second of these terms is an expression of broader meaningfulness of this relationship.

In the fuzzy framework, where the value of case i may be 0, 1 or a value between 0 and 1, fuzzy cardinality is defined in a similar manner to the crisp case. Fuzzy intersection between seeL ("low," i.e., good, seeing) and two (arbitrary) meteorological variables A and B, is defined as: $\sum_i \min(seeL_i, A_i, B_i)$. Note how this latter definition gives the crisp definition, if all variables happen to be crisp.

Conditional possibility is the name given to the generalized conditional probability. Let $\bigcap$ represent fuzzy intersection as defined above. We have:

$$poss(seeL \mid A, B) = \bigcap_i (seeL_i, A_i, B_i) / \bigcap_i (A_i, B_i)$$

$$= \sum_i (\min_i (seeL_i, A_i, B_i)) / \sum_i (\min_i (A_i, B_i))$$

This is the conditional possibility of good seeing, given A and B; i.e., the possibility of having good seeing when we are given these two meteorological variables.

To quantify associations of interest, conditional possibility will be used. This is the conditional possibility of good seeing on one or more meteorological variables. This is a fuzzy success rate, given a particular environment. A second measure will be the possibility of these meteorological variables arising in practice:

$$\bigcap_i (A_i, B_i) / n$$

where n is the total number of cases considered.

These definitions are used in the results shown in Table 3.6. Miyamoto [59] may be referred to for more discussion of fuzzy generalizations of similarity, conditional dependence and other concepts. Table 3.6 shows a selection of interesting conclusions derived from Figures 3.3 and 3.4. Some of these relationships are very good indeed. We see, for instance, that in autumn 1992, the meteorological variables s1L, s1dL and pH (respectively, low wind speed,

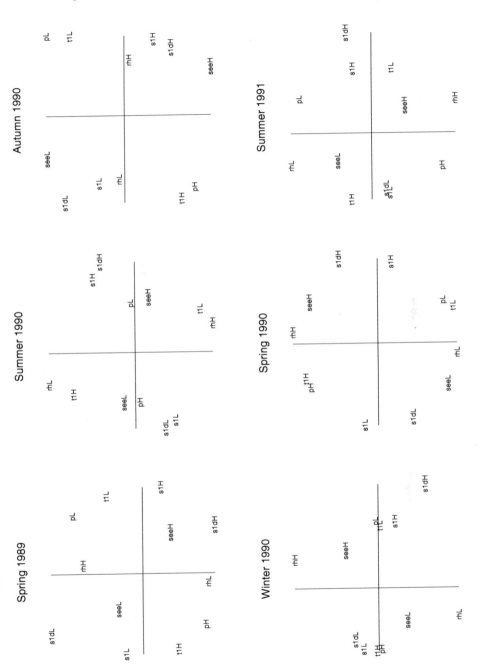

FIGURE 3.3

Principal planes of 11 correspondence analyses, carried out on data from 11 seasons between 1989 and 1992. The variables used are listed in Table 3.5. Continued in following figure.

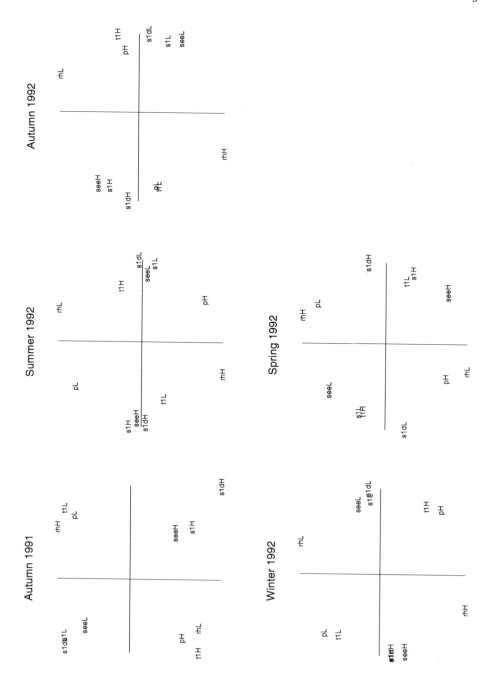

FIGURE 3.4

Continuation of analysis results of previous figure.

TABLE 3.6

Motivated by the proximities between variables in the 11 different correspondence analysis results shown in Figures 3.3 and 3.4, the above relationships were determined.

Season	No. of cases	Good seeing given ...	Conditional possibility of good seeing	Possibility of variables happening
Spr89	81	s1L	0.68	0.35
		s1dL	0.61	0.27
		s1L and s1dL	0.62	0.20
Sum89	77	pH	0.64	0.32
		s1dL	0.69	0.38
		s1L	0.68	0.35
		pH, s1dL and s1L	0.75	0.24
Aut90	48	rhL	0.76	0.35
Win90	69	rhL	0.63	0.32
		t1H	0.63	0.31
		pH	0.63	0.33
Spr90	47	s1L	0.58	0.29
		s1dL	0.67	0.36
		s1L and s1dL	0.67	0.27
Sum90	38	rhL	0.52	0.27
		t1H	0.62	0.31
		rhL and t1H	0.60	0.21
Aut91	61	s1L	0.61	0.29
		s1dL	0.70	0.41
		s1L and s1dL	0.60	0.25
Sum92	55	s1L	0.76	0.40
		s1dL	0.77	0.38
		pH	0.68	0.35
		s1L, s1dL and pH	0.88	0.24
Aut92	66	s1L	0.78	0.40
		s1dL	0.80	0.43
		pH	0.67	0.37
		s1L, s1dL and pH	0.90	0.31
Win92	44	s1L	0.89	0.46
		s1dL	0.87	0.42
		rhL	0.72	0.38
		s1L, s1dL and rhL	0.96	0.27
Spr92	69	s1L	0.74	0.38
		s1dL	0.68	0.33
		s1L and s1dL	0.72	0.29

low wind speed standard deviation, and high pressure) had a 90% association with good (i.e. "low") seeing; and that such a "constellation" of meteorological variables occurred 31% of the time. From Table 3.6, it is also seen that variables s1L and s1dL often co-occur with good seeing. Unfortunately, however, this is not always the case.

Within a season, it appears from Figures 3.3 and 3.4, and from Table 3.6, that there is information available which would allow good forecasts. This encourages looking further for the best operational approaches. Further discussion on this study can be found in [5, 6, 66]. Nonlinear regression was used, in the guise of a novel neural network approach, which embodied both time-varying and recurrent network links.

3.3 Assessment of Coding Methods

Clearly the use of coding methods such as doubling, complete disjunctive form and fuzzy coding, is partially forced upon the analyst. This is true in particular for handling mixed quantitative and qualitative data. It is also true when some form of variable weighting is required. Coding can also aid greatly in the interpretation of results.

In this section we will take a well-known data set and force these recodings on it. Our objective is solely to assess the outcome. We stress that we do this not to get better insight into the data (we have all the insight we need already) but simply to have a controlled assessment framework.

The Fisher iris data [42] consists of 150 observations, class 1 (observations 1 to 50) being clearly distinguished from the other two (observations 51 to 100, and 101 to 150) which are somewhat more confused. The data is quantitative or continuous, and is in 4-dimensional space. Figure 3.5 shows histograms of the 4 variables and Figure 3.6 shows a principal plane of this data with known class assignments indicated.

Doubling this data, using each variable's median as a threshold, is obviously going to degrade the information content available to us. Many projections overlap and observations in class 1, even, are confused with the other classes.

Figures 3.7 and 3.8 show results obtained for fuzzy coding. If α is a fraction, the two-pivot coding used as output codes is: $(1, 0), (\alpha, 1 - \alpha), (0, 1)$. The three-pivot coding used as output codes is: $(1, 0, 0), (\alpha, 1 - \alpha, 0), (0, \alpha, 1 - \alpha), (0, 0, 1)$. The pivots were taken as the 33rd and 67th percentiles in the first case; and the 25th, 50th and 75th percentiles (quartiles) in the second case. It does appear that Figure 3.8 shows greater differentiation of observations, consistent with the fact that it uses more information about the data compared to Figure 3.6.

Figure 3.9 is the result of recoding the Fisher data into a very high-dimension-

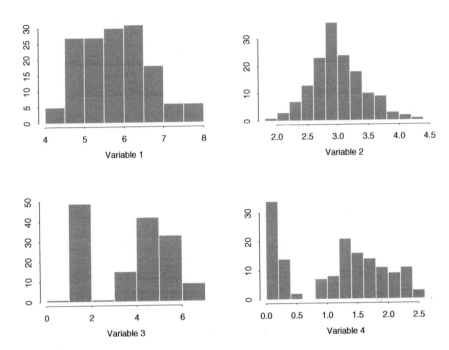

FIGURE 3.5
Histograms of four variables used in the Fisher iris data set.

al space. The four continuous real variable values were each recoded to one value of a 147-dimensional boolean vector. This mapping was done roughly by multiplication by 100, truncation and rounding. Each 147-valued observation vector thus contained exactly four 1-values. This implies that the observations were now equally weighted. Some of the 147 new variables were 0-valued for all observations. Removing these left 123-valued vectors, and the array to be analyzed was of dimensions 150 ×123. Figure 3.9 shows that class 1 is fairly clearly distanced from the others, but that there is some confusion between classes 2 and 3. The high-dimensional representation of the four-dimensional data was, as already mentioned, quite arbitrary. One could easily devise a more optimal mapping, given some optimality criterion.

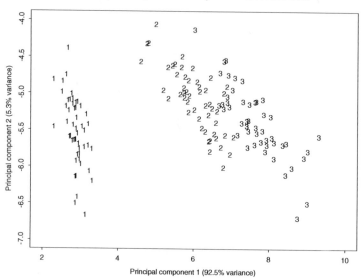

FIGURE 3.6

Principal components analysis of the Fisher iris data, showing a clearly separate class 1, and some confusion between classes 2 and 3.

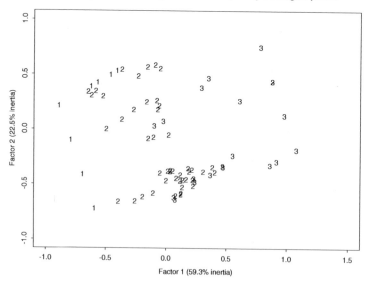

FIGURE 3.7

Principal factor of correspondence analysis of fuzzily-coded iris data. Two pivots or hinges were used for this coding.

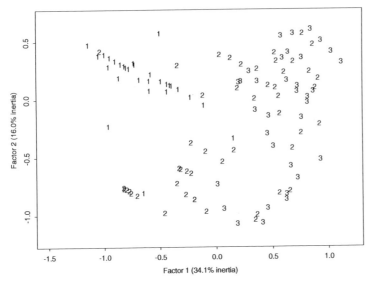

FIGURE 3.8

Principal factor of correspondence analysis of fuzzily-coded iris data. Three pivots were used for this coding.

The high-dimensional booleanized Fisher data was of interest to us for another reason also. In information retrieval, such high-dimensional spaces are common-place. The variables in such a case are index terms or keywords. Such document versus index term data is also often very sparse. Our booleanized Fisher data was also sparse, having precisely 4×150 one-values in an array of dimensions 150×123. Cluster analysis is sometimes used on the documents to facilitate retrieval of associated relevant documents. But sparse data mitigates against clustering since there may well be zero association between documents due to no common index terms. This is an argument for the use of dimensionality reduction, perhaps in conjunction with clustering.

A combined dimensionality reduction and clustering method is provided by the Kohonen self-organizing feature map [48, 65, 67]. The trained map can be used for information and resource discovery. Examples include Internet newsgroups [86] and the astronomical literature [72].

Figure 3.10 shows a Kohonen map of the original Fisher iris data. The user can trace a curve separating observation sequence numbers less than or equal to 50 (class 1), from 51 to 100 (class 2) and above 101 (class 3). Some observations may be superimposed, and thus not seen on this output. The zero values indicate map nodes or units with no assigned observations. The map of Figure 3.10, as for Figure 3.11, has 20×20 units. The number of epochs used in training was in both cases 133.

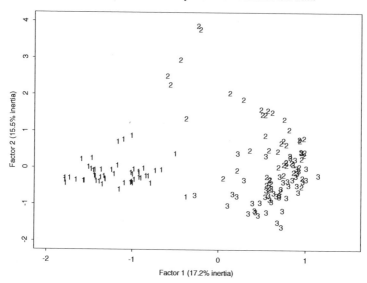

FIGURE 3.9

Principal factor of correspondence analysis of 123-dimensional booleanized iris data.

Figure 3.11 shows the Kohonen map resulting from the analysis of the 147-dimensional booleanized version of the Fisher data. The user in this case can demarcate class 1 observations. Classes 2 and 3 are more confused, even if large contiguous "islands" can be found for both of these classes. The result shown in Figure 3.11 is degraded compared to Figure 3.10. It is not badly degraded though insofar as sub-classes of classes 2 and 3 are found in adjacent and contiguous areas. We conclude from this that a method such as the Kohonen map method does a fairly good job of handling high-dimensional, sparse data. As mentioned earlier, a better high-dimensionally mapped version of the iris data (using binary classification trees, for instance) could well lead to even better fidelity of Kohonen output to the Kohonen analysis of the original data.

This assessment serves to indicate the increasing expressiveness, and the limitations, of the coding methods described here. These techniques are interpretation-friendly but may involve loss of information.

From Figure 3.5, variables 3 and 4 are multimodal, implying that analysis of correlations as implemented by principal components analysis is somewhat crude – no account is taken of the modes. The forms of coding looked at provide alternatives.

```
  0  70   0  82   0  61  94  58   0  42   0  39  43  48   0   7  25   0  23   0
  0   0   0   0   0   0   0  99   0   0   9   0   4  30   0   0  12   0   0   0
 63   0   0  81   0   0   0   0  80   0  46   0  31   0  36  50   8   0  41  38
  0   0   0   0   0  60   0   0  68   0  26  35   0   0   0   0  40  18   0   0
 54   0  90   0   0   0   0   0   0   0   0   0   0  27  32   0  29  28   0  47
  0   0   0   0  91  95   0   0  93   0  65   0   0   0   0  21  37   0  49   0
 88   0  73   0   0   0 100   0   0   0   0   0   0  44   0   0   0  11   0  33
120   0   0   0 107   0   0   0  72   0  97  96   0   0   0  22   0   0   0  34
135   0  84   0   0   0  56   0   0   0   0  89   0   0  45   0   0  17   0  15
  0   0 143   0   0   0   0   0  74   0   0   0   0   0   0   0   0   6   0  19
  0 114   0   0 127   0  79  64   0   0  85   0  62   0   0   0   0   0   0   0
147   0 122   0 124   0   0   0  92   0  67   0   0   0   0   0   0   0  98   0
112   0 115   0   0   0 134   0   0   0   0   0   0  86   0  52   0   0   0  75
  0   0   0   0 104   0   0   0   0 139 150   0   0   0   0   0   0   0  59   0
133   0   0   0   0   0 138   0 128   0   0   0  71   0  57   0  76   0   0  77
  0   0 113   0   0   0 148   0   0   0   0   0   0   0   0   0   0   0   0   0
130   0   0 140 125   0 146   0 105   0 116   0   0   0   0   0   0  87   0  66
108 131 103   0   0   0 142   0   0   0   0   0 111   0   0   0   0   0   0   0
123   0   0   0 110 121 141   0   0 137 149   0   0   0   0  78   0  53   0  51
119 136   0 132   0   0 144 145 101   0   0   0   0   0   0   0   0   0   0   0
```

FIGURE 3.10
Kohonen self-organizing feature map of the iris data.

```
109   0 127   0  91   0   0  77   0 134   0  87  53   0  66   0 145   0 110  58
104   0   0   0 119   0   0   0   0   0   0   0   0   0   0   0   0   0   0   0
124   0 138   0   0   0 123   0   0 115   0   0   0 140   0 141   0 129   0 133
128   0 117   0 132   0   0   0 131   0   0  73   0 142   0   0   0   0   0   0
150   0 148   0   0   0 114   0   0   0   0   0   0   0   0 135   0 137   0  17
  0   0   0   0 111   0 147   0  93   0  81   0   0   0   0   0   0   0   0   0
 92   0  14   0   0   0   0   0   0   0   0   0   0  22   0  82   0  32  16
  0   0  13   0  68 143   0 112   0 149   0  27   0   0   0   0   0   0   0   0
106   0  46   0   0   0   0   0   0   0   0   0   0  49   0   0  33   0  80
  0   0 113   0  84   0  83   0   0  31   0  21  40   0   0   0  10   0   0  61
146   0   0   0   0   0   0  70   0   0   0   0   0   0  28   0  35   0   0   0
136   0  78   0   0   0  99   0   0  47   0  29   0   0   0   0   0   0  94  44
  0   0 120   0 107   0  24   0  45   0   0   0   0   0  50   0  26   0   0   0
 85  69   0   0   0   0   0   0   0   0   1   0  23   0   5   0   0   0   0 121
  0   0  79  86   0   0   0  20   0  18   0   0   0   0   0   0   0   0 144   0   0
 52   0   0   0   0 118   0   0   0   0  41   0  34   0  48   0   0   0   0 116
 56   0  59  97   0   0   0  19   0   0   0   0   0   0   0   0  51   0   0   0
100   0   0   0   0  95   0   0   0  42   0   0   9   0  15   0   0   0   0 101
  0   0   0  65   0   0   0  54   0   0   0   0   0   0   0   0   0  64   0   0
 72   0  98   0  89  88   0  90   0  37  39  43   3  30   0  36   0  74   0  57
```

FIGURE 3.11
Kohonen map of the 147-dimensional booleanized version of the iris data.

3.4   The Personal Equation and Double Rescaling

Applying the Personal Equation

Frequently in opinion surveys a subject is asked to give a rating between 0 and 10 to a set of politicians, or to a set of statements: 0 to indicate total disapproval, 10 for total approval, and intermediate ratings to mark the degree of approval/disapproval. Faced with such a rating scale with several gradations, the subjects show a behavior which is only partially determined by their opinion on what is at issue. Independently of the issue involved, some subjects have a tendency to give extreme responses, either favorable or adverse, and others prefer to stick to the middle of the scale.

Experience has shown that the more freedom there is for a subject in his or her responses the more difficult it is to interpret them. On the other hand, a too rigid format (such as only two categories: "for" or "against") puts the subject so much ill at ease that we can no longer rely on the responses.

"The practice of correspondence analysis has however established that we gain by considering the mean of the scores attributed by a given subject as the zero-point of the scale adopted by him, in order to use this zero for rescaling the scores between −1 and +1" [22]. This is done using a formula known as personal equation, particular to each subject.

For each subject i, the rescaling between −1 and +1 of all the scores attributed by him or her is done by computing their mean (ave), maximum (max) and minimum (min). The scores are first centered by subtracting the mean from them. Then all the positive scores are divided by (max − ave); all the negative scores are divided by (ave − min); thus the scores given by the subject i vary from −1 to +1.

Now let $k(i, j)$ be a rescaled score; we code it across three categories by applying the formula:

```
if k(i,j) <= 0 then
        k(i,j+) = 0
        k(i,j=) = 1+k(i,j)
        k(i,j-) = k(i,j)
else
        k(i,j+) = k(i,j)
        k(i,j=) = 1-k(i,j)
        k(i,j-) = 0
endif
```

It is easy to recognize a barycentric principle in this coding, since the same result is achieved if we used the min, ave, max of each row i as the hinges for barycentrically coding all the scores in that row.

Further reading on the personal equation may be found in [13, 22, 53].

Double Rescaling

Some statements in a survey can receive a high degree of approval, resulting in high overall scores and some are generally strongly disapproved, resulting in poor overall scores. If, e.g., a proposition is in general scored low by all the subjects, the corresponding v^+ category will have a very low (or even zero) mass, with an ill-defined profile which might perturb the analysis. In order to overcome this drawback, we may, after rescaling the scores row by row, rescale them column by column. What this amounts to is this: If a proposition has been scored very low in general, any score which is above the average score for this question is considered as a sort of approval. In order to obviate any confusion that may result, the categories after this double rescaling are named v<, v≈, v>. This double rescaling is achieved by calculating, for the rescaled table (the table in which each row has been rescaled), the min, ave and max of each column, and using them as hinges for barycentrically coding the columns.

The relevance of this double rescaling is to produce balanced categories j> and j<, whether the entity j (e.g., proposition) is generally well scored or poorly scored (whereas the category j^- is as much heavier as j is poorly scored by the majority of the subjects). This is because here the mean point used for the second rescaling is adapted each time to j. Here, too, it is a barycentric coding.

It should however be borne in mind that the larger the number of transformations effected on the data, the more circumspect one should be. One of the ways of ensuring that the coding does not distort the data is to check the coherence of the results after each transformation.

Further reading on rescaling of rows and of columns may be found in [57].

3.5   Case Study: DNA Exon and Intron Junction Discrimination

The data used in this section is from L. Prechelt's neural network data set repository, Proben1 [74], and was used in [82]. More particularly relevant, it was used in [2, 27], and the following is a short introduction to those articles.

The nucleic acids found in chromosomes define the proteins produced in any organism, following a universal code. Briefly, the chromosome consists of molecules arranged in a double helix, and the signifying part of the chain is a sequence of bases that can be taken as a text written with four letters: T (thymine), A (adenine), C (cytosine) and G (guanine). Triplets, or codons, define an amino acid. For example, alanine is coded by one of the 4 triplets, GCT, GCA, GCC, GCG, each of which begin with GC. A sequence of triplets can be translated, through a complex physico-chemical process, into a sequence of amino acids, that is, a protein. In the long sequence of a chromosome, certain segments only that are copied from DNA (deoxyribonucleic acid) to messenger RNA (ribonucleic acid) govern, in the cytoplasm on the level of the ribosomes, the assembly of a protein. Other segments are eliminated during splicing of the RNA. In the DNA sequence there are marks of potential separation delimiting an exon, which serves as code for a protein, from an intron which is not used in a protein.

The data used consists of 3190 nucleotide sequences of 60 characters of the alphabet T, A, C, G. These sequences, derived from chromosomes of different primates, each have 20 triplets. The original database was "Primate splice-junction gene sequences (DNA) with associated imperfect domain theory" from Genbank 64.1 (genbank.bio.net). If the junction of the 10th and 11th triplets mark an exon-intron ("donor") junction then the sequence is labeled EI. If the junction of the 10th and 11th triplets mark an intron-exon ("acceptor") junction then the sequence is labeled IE. Finally, if the junction of the 10th and 11th triplets does not mark either exon-intron nor intron-exon, then the sequence is labeled N. Approximately, EI and IE junctions comprise 25% each of the sequences, and N junctions comprise 50% of the sequences. An example follows.

```
EI, CCAGCTGCATCACAGGAGGCCAGCGAGCAGGTCTGTTCCAAGGGCCTTCGAGCCAGTCTG
IE, CCCGAATGATCTCCAGCATTCTGTGCCTAGCTGCTGATCGCCTACAAGCCAGCCCCTGGC
N,  ATGATGTTTGCCTCCGCCCTGCCTGCTCTGCTGGTCTTCATCCTCATATTCCTGGAGTCT
```

In the 60 characters of each sequence, [27] investigates the distribution of pairs of characters across the 30–31 character boundary; and the nearby boundaries, 32–33 and 28–29. In all cases, the EI, IE and N class numbers are reported on, for the two characters at these boundaries. This is one way to address the question as to how to carry out discrimination between the three classes, viz. EI, IE and N. Such discrimination is based on consecutive pairs of characters.

Extending this to a multivariate setting, we take 12 bases, or 4 triplets, around the crucial 10th/11th triplet junction. Using complete disjunctive coding, each basis is mapped onto one of $(0,0,0,1)$, $(0,0,1,0)$, $(0,1,0,0)$, $(1,0,0,0)$. So the 12 bases in each sequence therefore define a boolean vector of length 48 values, in all. The number of sequences used is 3175 (where a small number was filtered out due to missing data). Specifically we used file gene1.dt [74], recoding it into complete disjunctive form, in the following.

The coded bases give 48 values, and the 3 classes (EI, IE, N) give 3 values: in total the input data crosses 3175 nucleotide sequences by 51 boolean-coded values. Call this data matrix X, of dimensions 3175×51. As seen earlier in this chapter and in chapter 2, the Burt table is the matrix XX^t of dimensions 51×51 (where t denotes matrix transpose). This data table is analyzed as a principal table, and the table X^t is conjoined as a set of 3175 supplementary columns. This will allow us to find the factor space projections and then display the sequences, by basing the analysis on the chosen nucleotide bases and their EI, IE or N classes. It is a neat way – both in effectiveness and efficiency – to carry out such an analysis.

Figure 3.12 shows the output display of the principal factor plane, where a relatively clear set of zones corresponding to the classes can be distinguished. The frontiers between junction classes indicates some overlap, but the zones themselves are quite clearly demarcated. Schematically, the cloud of 3175 sequences can be considered as a rectangle, with trapezium regions for EI and IE. Such a view of the data can be used for discrimination purposes.

In [27], the analysis continues with the following.

1. Hierarchical clustering of the 48 response categories corresponding to selected bases; and the projections of the cluster members on the principal factor plane.

2. Hierarchical clustering of the over three thousand sequences.

3. Nearest class mean discrimination, to allow for assignment of each sequence to its most appropriate class. Better results are obtained for assignment to principal columns, compared to supplementary rows. Combining these by taking the average of two different definitions of class center gives rise to the best results of all.

4. A closer discrimination-oriented look is taken at how bases right at the crucial location of the 30th and 31st ranks are the most influential. This is followed by a focus on four classes of sequence related to guanine, G (which are: GG, Gx, xG, xx, where x denotes a base other than guanine).

5. Finally all of the foregoing analyses were undertaken on the full data set. These analyses are undertaken anew on two-thirds of the sequences taken as the training set. The remainder of the data comprises the unseen, or test, set.

The principle of the nearest mean classifier used here is simple and well-known. Its success depends very much on the Euclidean space in which this classification is carried out. Therefore everything is underpinned by the multidimensional data analysis – including data coding – that has served to construct this space.

In [2], the analysis is extended from the 12 bases or 4 triplets, to the entire sequence. It is confirmed that the characteristic aspects of the intron are the most revealing; and that there is nothing found to characterize well the negative junction modality, N. A further conclusion of this work is that assignment to the best class should be based on a carefully – iteratively refined – subset of the data. A final viewpoint expressed is that the discrete logic inherent in classification, or for that matter in projections onto the "fixed logic" of orthogonal axes, will be inferior in the longer term to a spatial representation. It is the latter, the spatial representation, that coordinates the multiple facets of the real, observed structure in a global description.

Supplementary, 3175 sequences; / = EI, \ = IE, – = N

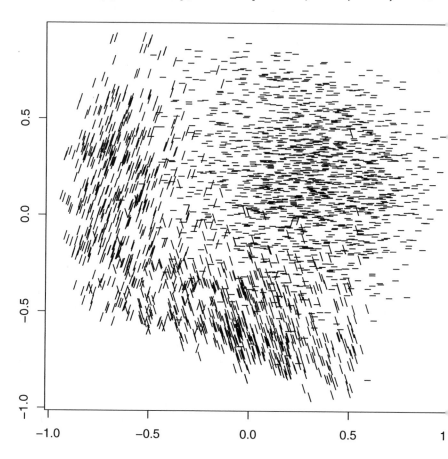

FIGURE 3.12

First two factors of correspondence analysis of exon/intron junction data.
3175 nucleotide sequences shown. The symbols used indicate the corresponding junction type, EI, IE or N.

Software Used

The R software used for producing Figure 3.12 was as follows.

```
# g is gene1.dt in complete disjunctive form
g <- read.table("c:/proben1/proben1/gene/gene1.new")
nrow(g); ncol(g)
[1] 3175
[1] 243

# Select the bases of interest
gx <- g[,97:144]

# Form Burt table
gxg <- t(as.matrix(gx)) %*% as.matrix(gx)

# Carry out both principal and supplementary analyses
gxgcs <- caSuppCol(gxg, t(as.matrix(g[,97:144])))

# Get identifiers for classes EI, IE, N
cl1 <- g[,241]==1
cl2 <- g[,242]==1
cl3 <- g[,243]==1

# Plot the data
plot(gxgcs[,1],gxgcs[,2],type="n",xlab="",ylab="")
text(gxgcs[cl1,1],gxgcs[cl1,2],"/")
text(gxgcs[cl2,1],gxgcs[cl2,2],"\\")
text(gxgcs[cl3,1],gxgcs[cl3,2],"-")
title("Supplementary, 3175 sequences; / = EI, \\ = IE, - = N")
```

3.6 Conclusions on Coding

An extensive suite of programs for handling the various coding situations considered here has been documented [28]. Case studies associated with this work have included use of "India SCAN 1996" data, which is part of a large international study of social climate produced using the "DYG SCAN" methodology of the social and marketing research company DYG Inc. Another case study has dealt with 20 years' political survey data from former East Germany and former West Germany. This is part of the "Politbarometer" longitudinal study of political and social attitudes and perspectives, produced by the Zentralarchiv für empirische Sozialforschung at the University of Cologne.

The motivation and justification for the study of the topics dealt with here come from practical and real world issues. Real data analysis never has the luxury of "clean" data. Instead, data are imprecise, partially existent and mixed in terms of various traditional data categorizations (qualitative and quantitative, text and numeric, macro and micro, and so on). Coding is part and parcel of the data analysis *process*.

Our discussion of recoding for correspondence analysis has provided an overview of a rich and extensively developed set of techniques. Such recoding practices are of importance for other related data analysis methods also – other multivariate analysis methods, decision-support systems, data mining, neural networks, and data visualization and display methods.

3.7 Java Software

Software discussed here, and data sets, are available on the web at: www.correspondances.info .

The Java sources are available for download as `java-source.tar`. The files can be extracted using PKUnzip or some similar Windows utility, or `tar xvf java-source.tar` on a Unix system. Here are the programs:

`CholeskyDecomposition.java`	`Maths.java`
`CorrAnal.java`	`Matrix.java`
`DataAnalysis.java`	`PrintUtilities.java`
`EigenvalueDecomposition.java`	`QRDecomposition.java`
`Header.java`	`ResultCA.java`
`HierClass.java`	`ResultHC.java`
`Imatrix.java`	`ResultInterpret.java`
`Interpret.java`	`ResultSVD.java`
`LUDecomposition.java`	`SingularValueDecomposition.java`
`ManageOutput.java`	

Quite a few of these programs are from: JAMA : A Java Matrix Package (National Institute of Standards and Technology), http://math.nist.gov/ javanumerics/jama . The programs were set up by us to allow everything here to be run without the need to download and install programs from elsewhere.

Assuming you have the Java environment, JDK or Java Development Kit, on your system, you will be able to compile these programs. If you have the `javac` program this will be used as follows. Assume you have placed all the above `*.java` programs in a directory or folder called `java-sources`. Then, create a similar directory called `java-classes`. In directory `java-sources`, use the command: `javac *.java -d ../java-classes` (for Unix, or for Windows: `javac *.java -d ..\java-classes`). That will produce the corresponding `*.class` files in the directory `java-classes`.

Java documentation, in HTML, can be created by using the `javadoc` command. In the source file directory, use the following: `javadoc *.java -d java-doc`. This creates the HTML documentation files in directory `java-doc`. Using the `index` entry file, read this documentation with your web browser.

For the case where the Java Development Kit is not on your system, these class files have already been created. They are in `java-classes.tar`. Extract these in a directory, e.g., `java-classes`. To run them you need the Java Runtime Environment, or JRE, on your system. In particular you will want the command `java` to be known, i.e., accessible to you, on a known path.

If you have any difficulty in getting either JDK or JRE, you will find what you need at http://www.javasoft.com.

To simplify usage, and to bypass discussion of specifying paths, let us now run the Java software. It is assumed that you have at least JRE on your system. Go to the directory containing the classes. Do: `java DataAnalysis`. That is all: that is the entry point to the program package. Do this as a command-line instruction. A pop-up window will now prompt for the input data file.

The following are the objectives of the Java software:

- To provide functionality for the core algorithms – correspondence analysis as factor analysis, hierarchical clustering, and the VACOR and FACOR (respectively, variables versus clusters, and factors versus clusters) interpretational tools.

- Therefore to provide software components for larger, dedicated, customized or embedded systems.

- To have all necessary additional linear algebra functionality integrated with these algorithms in a single ready-to-use package.

- To cater for cross-platform use, given that Java sources (*.java) or byte code (*.class) are portable in this way. (The R environment is supported also for all common platforms.)

- To provide very basic input and output support, catering respectively for varied input data formats, or production quality graphical output. The R infrastructure addresses far better all aspects of input and output.

3.7.1 Running the Java Software

Using data file `phosphates.dat` we have a 14×8 data table. The first line is a title.

Then we have: numbers of rows, columns; an indication of no supplementary elements; output for 7 factors is requested; in the interpretation aids, a 4-cluster solution on the rows is to be used; and an 8-cluster solution on the columns. Finally on this line we have the column identifiers.

Then comes the data with row identifiers leading each row.

```
Commerce mondial des phosphates
 14 8 SUPNO 7 4 8   eBL    eUS    eJR    eMR    eSN    eTG    eTN  eCC
iBL      0  1305     0    3573     25    500    110    293
iCA      2  8335     0       0      0      8      0      0
iFR   1311  2691    70    4891   1484   2526   1697      4
iDL   1322  3808     0    1445    261    200    288   1442
iIT     42  1883   194    2881     67    195    493      0
iJP      0  4426   522    1540    239     93      0      0
iNL    299  1484     0    1853    249   1584     59     84
iSP     20   339     0    5073      2     85     11      0
iUK    122   645     2    2852    971     36    197      0
iIN      0  2559   996    1149    218      0    134      0
iBR     29  4918     0    1398     15      0    241      0
iPL      0  1284   271    3311     33    540    548   1996
iRM      0   768   483    1541      6    138     22   1206
iEE      2   201   136    1398      0      0    333   5533
```

The Java software requires exactly the right format for the input data. It will fail otherwise. A set of eight small (overlapping) windows is created, containing the output. Some of these will be looked at in turn.

The first window contains the factor analysis results. Below we cut the output at four factors.

```
phosphates.dat
Commerce mondial des phosphates
trace  :  1.1124
Lambda, rates and cumulative values are scaled by 10^-4
rank   :     1     2     3     4     5     6     7
lambda :  4505  3161  1524   819   575   335   206
rates  :  4050  2842  1370   736   517   301   185
cumul  :  4050  6891  8261  8997  9514  9815 10000
```

```
|IDNI| QLT WTS INR|  F#1 CO2 CTR|  F#2 CO2 CTR|  F#3 CO2 CTR|  F#4 CO2 CTR|
| iBL|1000  60  22|  -20   1   0|  405 390  31|  359 306  50|  294 206  63|
| iCA|1000  86 139| -786 342 117|-1038 595 291| -130   9   9|  266  39  74|
| iFR|1000 150 102| -208  57  14|  649 556 200| -498 328 245| -153  31  43|
| iDL|1000  90  50|   60   6   1| -242  94  17| -516 427 157|  108  19  13|
| iIT|1000  59  15| -226 180   7|  209 155   8|  318 357  39|   52   9   2|
| iJP|1000  70  37| -514 450  41| -473 381  50|  223  84  23| -190  62  31|
| iNL|1000  58  51| -242  59   7|  523 279  50| -444 200  74|    6   0   0|
| iSP|1000  57  77|    9   0   0|  767 389 105|  823 449 252|  436 126 131|
| iUK|1000  49  56| -156  19   3|  716 410  80|  192  29  12|  -47   2   1|
| iIN|1000  52  63| -398 117  18| -411 124  28|  472 164  76| -891 585 502|
| iBR|1000  68  44| -569 444  49| -554 422  66|   53   4   1|  270 100  60|
| iPL|1000  82  29|  559 802  57|   95  23   2|  146  55  11|    4   0   0|
| iRM|1000  43  30|  643 525  39| -179  41   4|  380 184  40| -381 184  76|
| iEE|1000  78 284| 1933 922 647| -523  67  67| -129   4   8|   56   1   3|
```

IDNJ	QLT	WTS	INR	F#1	CO2	CTR	F#2	CO2	CTR	F#3	CO2	CTR	F#4	CO2	CTR
eBL	1000	32	67	-146	9	2	440	84	20	-1151	575	281	-47	1	1
eUS	1000	355	201	-528	443	220	-585	543	385	-50	4	6	76	9	25
eJR	1000	27	80	1	0	0	-466	67	19	785	190	111	-1519	713	773
eMR	1000	337	122	45	5	2	492	602	259	371	342	304	134	45	74
eSN	1000	37	64	-299	46	7	768	306	68	-432	97	45	-460	109	94
eTG	1000	61	95	-156	14	3	822	387	129	-692	275	190	-136	11	14
eTN	1000	42	32	101	12	1	450	239	27	-407	195	46	-184	40	17
eCC	1000	108	339	1785	913	765	-523	78	94	-155	7	17	34	0	2

Windows 2, 3 and 4 show the plane of factors 1 and 2; 1 and 3; and 2 and 3. These are displayed in a very simple line-mode way.

Window 5 displays the hierarchical clustering, again using a simple line-mode way of doing this.

```
Inertias and level indices are calculated in
the space defined by the 7 axes used for the CAH.
The sum of level indices is 1.1124
Rates T are scaled by 10^-4
c      27   26   25   24   23   22   21   20   19   18   17   16   15
car    14   13    9    7    3    4    4    2    3    3    2    2    2
T    3079 2444 1049  938  628  535  391  229  224  214   96   94   76
A     iEE   22   20   21  iDL  iIN  iUK  iFR  iCA  iSP  iPL  iJP  iBL
B      26   25   24   23   17   19   18  iNL   16   15  iRM  iBR  iIT
```

```
Plot of T against c
Scale 0..20 used for ordinate
|                                        XX
|                                        XX
|                                        XX
|                                        XX
|                                     XX XX
|                                     XX XX
|                                     XX XX
|                                     XX XX
|                                     XX XX
|                                     XX XX
|                                     XX XX
|                                     XX XX
|                                     XX XX
|                               XX XX XX XX
|                               XX XX XX XX
|                            XX XX XX XX XX
|                         XX XX XX XX XX XX
|                      XX XX XX XX XX XX XX
|                XX XX XX XX XX XX XX XX XX
+----------------------------------------
    15 16 17 18 19 20 21 22 23 24 25 26 27
```

```
Representation of the hierarchy.
Classes are labeled n+1 to 2*n-1.
   27   27   27    27   27    27    27   27    27   27   27   27   27   27
```

```
14   26  26  26  26  26  26  26  26  26  26  26  26  26
14   22  22  22  22  25  25  25  25  25  25  25  25  25
14   22  22  22  22  20  20  24  24  24  24  24  24  24
14   22  22  22  22  20  20  21  21  21  21  23  23  23
14   22  22  22  22  20  20  21  21  21  21   4  17  17
14   10  19  19  19  20  20  21  21  21  21   4  17  17
14   10  19  19  19  20  20   9  18  18  18   4  17  17
14   10  19  19  19   3   7   9  18  18  18   4  17  17
14   10   2  16  16   3   7   9  18  18  18   4  17  17
14   10   2  16  16   3   7   9   8  15  15   4  17  17
14   10   2  16  16   3   7   9   8  15  15   4  12  13
14   10   2   6  11   3   7   9   8  15  15   4  12  13
14   10   2   6  11   3   7   9   8   1   5   4  12  13
iEE iIN iCA iJP iBR iFR iNL iUK iSP iBL iIT iDL iPL iRM
```

```
Number of clusters: 4
Cluster contents follow.

Cluster i14: iEE
Cluster i20: iFR iNL
Cluster i22: iIN iCA iJP iBR
Cluster i24: iUK iSP iBL iIT iDL iPL iRM
```

Window 6 gives the FACOR (i.e., factors and clusters) interpretational output. We cut this here to show the results for 3 factors.

```
---------------------------------------------------------------------------
|CLAS ELDR YNGR| QLT WTS INR|  F#1 CO2 CTR|  F#2 CO2 CTR|  F#3 CO2 CTR|
---------------------------------------------------------------------------
Representation on the factorial axes of the 3 selected nodes
   27   14   26|  01000   0|    0   0   0|    0   0   0|    0   0   0|
   26   22   25|1000 922  24| -163 922  55|   44  67   6|   11   4   1|
   25   20   24|1000 647  70|   18   3   0|  342 970 240|  -35  10   5|
Representation on the factorial axes of the 4 classes of the selected part.
   14    0    0|1000   78 284| 1933 922 647| -523  67  67| -129   4   8|
   20    3    7|1000  208 130| -217  68  22|  614 541 248| -483 335 319|
   22   10   19|1000  275 198| -590 435 213| -657 538 375|  118  17  25|
   24   21   23|1000  439  45|  130 148  16|  213 400  63|  178 278  91|
---------------------------------------------------------------------------

---------------------------------------------------------------------------
|CDIP ELDR YNGR| QLT WTS IND|  D#1 COD CTD|  D#2 COD CTD|  D#3 COD CTD|
---------------------------------------------------------------------------
Representation on the factorial axes of the 3 selected dipoles
   27   14   26|10001000 308| 2097 922 701| -567  67  73| -139   4   9|
   26   22   25|1000 922 244| -608 263 159| -999 709 609|  153  17  30|
   25   20   24|1000 647 105| -347 146  38|  401 194  72| -661 528 405|
---------------------------------------------------------------------------
```

Window 7 gives the VACOR (i.e., variables and clusters) interpretational output. We cut this here to show the results for just the first 3 variables.

```
Clustering to be used for aggregating variables.
The sum of level indices is 1.1124
Rates T are scaled by 10^-4
c    15   14   13   12   11   10    9
```

```
car      8    7    5    2    4    3    2
T     3805 2930 1251  834  502  388  290
A      eCC   12  eMR  eUS  eTG  eBL  eSN
B       14   13   11  eJR   10    9  eTN
```

```
Plot of T against c
Scale 0..20 used for ordinate
|                      XX
|                      XX
|                      XX
|                      XX
|                   XX XX
|                   XX XX
|                   XX XX
|                   XX XX
|                   XX XX
|                   XX XX
|                   XX XX
|                   XX XX
|                   XX XX
|                XX XX XX
|                XX XX XX
|             XX XX XX XX
|             XX XX XX XX
|       XX XX XX XX XX XX
|    XX XX XX XX XX XX XX
+----------------------
      9 10 11 12 13 14 15
```

```
Representation of the hierarchy.
Classes are labeled n+1 to 2*n-1.
  15   15   15   15   15   15   15
   8   14   14   14   14   14   14
   8   12   12   13   13   13   13
   8   12   12    4   11   11   11
   8    2    3    4   11   11   11
   8    2    3    4    6   10   10
   8    2    3    4    6    1    9
   8    2    3    4    6    1    5
 eCC  eUS  eJR  eMR  eTG  eBL  eSN  eTN
```

```
Number of clusters: 8
Cluster contents follow.

Cluster j1: eBL
Cluster j2: eUS
Cluster j3: eJR
Cluster j4: eMR
Cluster j5: eSN
Cluster j6: eTG
Cluster j7: eTN
Cluster j8: eCC
```

```
Diagonalization in the context of variable aggregation.
phosphates.dat
Commerce mondial des phosphates
trace  :   1.1124
Lambda, rates and cumulative values are scaled by 10^-4
rank   :    1     2     3     4     5     6     7
lambda :  4505  3161  1524   819   575   335   206
rates  :  4050  2842  1370   736   517   301   185
cumul  :  4050  6891  8261  8997  9514  9815 10000
```

CLAS	ELDR	YNGR	QLT	WTS	INR	eBL	CO2	CTR	eUS	CO2	CTR	eJR	CO2	CTR

Representation of the 3 nodes on the set J

CLAS	ELDR	YNGR	QLT	WTS	INR	eBL	CO2	CTR	eUS	CO2	CTR	eJR	CO2	CTR
27	14	26	1000	1000	0	32	0	0	355	0	0	27	0	0
26	22	25	1000	922	24	35	8	35	383	75	61	28	1	7
25	20	24	1000	647	70	49	75	965	225	395	939	16	38	993

Representation of the 4 classes on the set J

CLAS	ELDR	YNGR	QLT	WTS	INR	eBL	CO2	CTR	eUS	CO2	CTR	eJR	CO2	CTR
14	0	0	1000	78	284	0	8	99	26	75	133	18	1	19
20	3	7	1000	208	130	79	99	568	206	90	73	3	30	320
22	10	19	1000	275	198	1	37	329	755	560	692	57	39	628
24	21	23	1000	439	45	35	2	4	234	362	101	22	9	32

CDIP	ELDR	YNGR	QLT	WTS	IND	eBL	COD	CTD	eUS	COD	CTD	eJR	COD	CTD

Representation of the 3 dipoles on the set J

CDIP	ELDR	YNGR	QLT	WTS	IND	eBL	COD	CTD	eUS	COD	CTD	eJR	COD	CTD
27	14	26	1000	1000	308	-35	8	107	-357	75	144	-10	1	21
26	22	25	1000	922	244	-48	51	553	529	560	854	40	42	847
25	20	24	1000	647	105	44	73	340	-28	3	2	-19	15	133

The final window summarizes the analysis. With the exception of this last window, all previous windows can be closed. But closing the last window terminates the analysis. Control is returned in the command prompt window which was used to start the analysis, by giving the java DataAnalysis command.

In the data sets provided on the book's web site there are examples of supplementary rows and columns. Respectively a final column, or a final row, of 1s and 0s must be provided. An indication of supplementary elements must also be provided on the parameter line following the title line.

4

Examples and Case Studies

4.1   Introduction to Analysis of Size and Shape

The chapter comprises three morphometric and biometric studies, followed by two economic and financial studies.

First, prehistoric goblets are used, with the analysis addressing issues related to shape and size. Also with a clustering or typology orientation, canine teeth are used in the second study. The third study describes 150 human skulls, discovered in Egypt, in ancient tombs in the region of Thebes. Using just 4 craniometric measurements here, we relate their properties to the chronological time line of 4000 BC up to 150 AD.

Chronology plays a central role in the fourth study, relating to world commerce in phosphates between 1973 and 1980. In the fifth study, in innovative work, we show how correspondence analysis and cluster analysis can be used for financial time series analysis, including both modeling and forecasting.

4.1.1   Morphometry of Prehistoric Thai Goblets

The shape of 25 prehistoric goblets (a drinking cup without handles but with a stem) [26, 55] is described by six measurements, $\{Wo, Wg, Ht, Ws, Wn, Hs\}$, with:

Wo = width, or diameter, of the opening at the top;

Wg = maximum width of the globe;

Ht = total height of the goblet;

Ws = width, or diameter, of the stem;

Wn = width of the stem at its top extremity, or neck, by which it is attached to the cup;

Hs = height of the stem; and

Hg = height, from the top of the stem to the horizontal plane of the top opening; this height is the difference, $Ht - Hs$, between the total height and the height of the stem.

The data set is available on the web at www.correspondances.info. For four specimens denoted by B, V, W, X, we have $Wg = Wo$, which means that at the opening the shape is that of a regular cup.

25 goblets, 6 attributes, 1 supplementary attribute

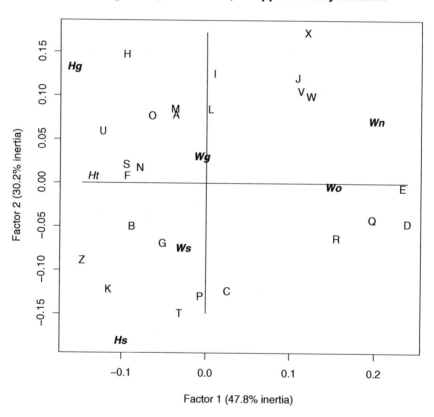

FIGURE 4.1

Factors 1 and 2 projections of 25 goblets, and 6 attributes. One attribute is supplementary.

TABLE 4.1

Thai goblet data analyzed: first three rows shown (see web site for remainder of data).

	Wo	Wg	Ht	Ws	Wn	Hs	Hg
A	13	21	23	14	7	8	15
B	14	14	24	19	5	9	15
C	19	23	24	20	6	12	12

The initial table cross-classifies the set I of 25 goblets (each of which is denoted by an upper case letter) with the set of six measurements $\{Wo, Wg, Ht, Ws, Wn, Hs\}$. In the table analyzed, shown in Table 4.1, both Hs and the difference Hg are used, while Ht is supplementary.

Correspondence analysis of the table of measurements is now discussed. The percentages of inertia explained by the top-ranked factors are 40%, 30%, 13%, 7%, 2%. Axis 1 therefore accounts for nearly one half of the total inertia. We have displayed the 1,2 plane in Figure 4.1; but, in the interpretation, we shall also discuss axis 3.

The total height, Ht, is perfectly correlated with axis 1 (see listing which follows: we are referring to the value of 962 thousandths in column CO2 for factor 1). This implies that like Ht, which is the barycenter of $\{Hs, Hg\}$, the barycenter of the 4 other principal measurements, viz. the widths $\{Wo, Wg, Ws, Wn\}$, is also approximately on axis 1, but on the side $F1 > 0$.

More precisely, $\{Wg, Ws\}$, maximum width of the globe and of the stem, are near the origin on the half-axis $F1 < 0$: the contrast is therefore between $\{Hs, Hg\}$ and $\{Wo, Wn\}$. On the side, $F1 > 0$, we find the slender shapes with a closed up globe (Wo small) resting on a tapering stem (Wn small); and, on the side $F1 < 0$, widening cups, on a stem whose shape borders on the cylindrical.

Thai goblet data, factor projections, contributions and correlations.

```
|SYMJ| QLT WTS INR|  F 1 CO2 CTR|  F 2  CO2 CTR|  F 3 CO2  CTR|  F 4 CO2 CTR|

|  Wo| 996 183 208| -150 717 313|   4    0   0| -91 262 414|  23  17  52|
|  Wg| 930 246  49|   6   6   1| -29 155  25| -10  19   7| -65 750 550|
|  Ws| 914 201  77|  27  70  11|  74 520 133|  30  87  51|  50 236 272|
|  Wn| 993  88 214| -201 598 268| -71  75  53| 147 320 519|   0   0   0|
|  Hs| 966 112 183| 101 224  86| 181 719 435|  14   4   6| -29  19  51|
|  Hg| 994 170 268| 159 575 322| -132 398 353|  -8   1   3|  29  19  75|
supplementary element
|  Ht| 967 282 194| 136 962 391|  -8   3   2|   1   0   0|   6   2   5|
```

Axis 2 is created by the contrast between Hs and Hg: on the side ($F2 < 0$) on which Hs lies, the height of the stem approximates, or even equals, that of the globe; on the side ($F2 > 0$), the height of the stem is only one third of the depth of the cup.

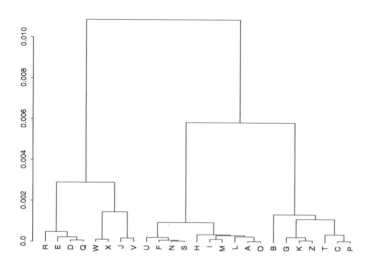

FIGURE 4.2

Hierarchical clustering of 25 goblets.

The goblet X will be examined in greater detail. With $F1 > 0$, it is a real cup, and not at all a wine glass. Its stem, small in height relative to all other measurements, conforms with $F2 > 0$. But overall, the width of the opening Wo is relatively large, while Wn, neck of the stem, is small. In the 1,2 plane, X is closer to Wn than to Wo. The inconsistencies are resolved if we consider axis 3, created exclusively by the contrast between Wn and Wo. On this axis, X, associated with Wn, is on the opposite side of Wo.

Cluster analysis more clearly explains this aspect of the shapes: Figure 4.2, followed by Figure 4.3.

FACOR output, as follows, provides information on factors for the 8-cluster partition of the goblets.

```
Number of clusters: 8
Cluster contents follow.

Cluster i2:  B
Cluster i32: W X
Cluster i34: J V
Cluster i36: U F N S
Cluster i37: G K Z
Cluster i40: H I M L A O
Cluster i41: T C P
Cluster i42: R E D Q

i40 + i36 --> i43
i2  + i37 --> i44
```

Hierarchical clustering of 6 goblet attributes

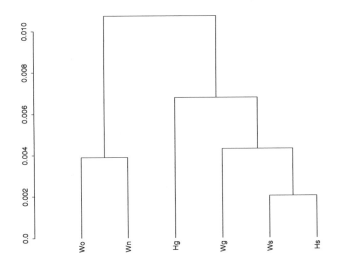

FIGURE 4.3

Hierarchical clustering of 6 goblet attributes.

```
i44 + i41 --> i45
i34 + i32 --> i46
i46 + i42 --> i47
i43 + i45 --> i48
i47 + i48 --> i49
```

CLAS	ELDR	YNGR	QLT	WTS	INR	F#1	CO2	CTR	F#2	CO2	CTR	F#3	CO2	CTR
Representation on the factorial axes of the 7 selected nodes														
49	47	48	0	1000	0	0	0	0	0	0	0	0	0	0
48	43	45	1000	740	102	61	990	211	5	7	2	2	1	1
47	42	46	1000	260	290	-175	990	601	-15	7	7	-5	1	2
46	32	34	1000	87	113	-114	360	85	-123	420	157	-86	203	174
45	2	44	1000	313	170	58	223	79	106	744	419	-20	27	35
44	37	41	1000	271	162	53	171	58	115	792	425	-20	24	30
43	36	40	1000	427	142	64	444	132	-68	508	239	18	35	38
Representation on the factorial axes of the 8 classes of the selected part.														
2	0	0	1000	42	57	89	209	25	50	65	12	-22	13	6
32	23	24	1000	34	112	-122	163	38	-137	207	77	-238	625	531
34	10	22	1000	52	53	-109	424	47	-114	463	82	14	7	3
36	21	30	1000	181	71	98	882	131	-27	66	15	24	51	27
37	7	31	1000	130	97	107	553	112	94	425	136	14	10	7
40	8	39	1000	247	105	39	128	28	-99	829	289	14	16	13
41	20	28	1000	141	106	4	1	0	134	861	302	-52	129	104
42	18	35	1000	173	281	-206	938	552	39	34	32	35	28	59

The listing FACOR shows that the topmost division is perfectly correlated with axis 1: $i48$, on the side $F1 > 0$ (slender shapes) contrasts with $i47$ on the side $F1 < 0$. Similarly, class 48 is split exactly along axis 2: $F2(i43) < 0$, low height of the stem; $F2(i45) > 0$, height of the stem equal to the depth of the cup.

The division of $i47$ into $i46$ and $\{R, E, Q, D\}$ takes place in the plane (2,3), mainly in the direction of axis 2: $\{R, E, Q, D\}$ is a class concentrated around its center, close to axis 1; whereas for $i46$, $F1$, $F2$ and $F3$ are negative. The last class splits into $\{V, J\}$ and $\{W, X\}$; the division is perfectly correlated with axis 3. For the goblets $\{W, X\}$, which are almost identical to each other, we have $F3 < 0$; Wn is small compared to Wo, the stem is negligible in all its dimensions. For $\{V, J\}$, the stem is placed under the cup as a sort of flattened or sawn-off cone of non-negligible width compared to the width of the opening, Wo.

Manly [55] considers the extent to which differences between goblets are due to shape rather than size, and suggests that the size differences be removed by dividing each of the measurements for a goblet by either the total height, Ht, or the sum of all our measurements. This is precisely what takes place automatically in correspondence analysis. The profile of a goblet is nothing other than the set of all its measurements divided by the total of all the measurements.

It turns out that all the goblets are of equivalent size, with the exception of $\{V, W, X\}$, and perhaps J. With these exceptions, their masses are of the same order. Looking at the column WTS in the correspondence analysis listing of the Thai goblet projections, contributions and correlations above (column 3) it is seen that the lowest masses are 17, 18, 22, 31 (W, X, V, J), while all other masses are between 33 and 51; and (W, X, V, J) is precisely the class $i46$. The conclusion of Benzécri and Gopalan [26] is therefore: the analysis of the shapes (or, in other words, correspondence analysis) suffices for acquiring an overall view of the diversity of the goblets.

4.1.2   Software Used

In R, we used the following code for the analysis of the goblets. The Windows syntax has to be changed a little (forward slashes to back slashes) for a Unix system.

```
# Goblets data array, Ht is at barycenter of Hs and Hg; Ht is
# supplementary.  What follows is self-explanatory.

gobelets <- read.table("c:/gobelets.dat")

nrow(gobelets); ncol(gobelets)
# Shows that there are 25 rows and 7 columns.
```

```
source("c:/ca.r")
source("c:/caSuppCol.r")

# Correspondence analysis: factor analysis on main elements.
gobCA <- ca(gobelets[,-3])

# We get the following:
# Eigenvalues follow (trivial first eigenvalue removed).
# 0.01325997 0.008382031 0.003654304 0.001865742 0.0005574158
# Eigenvalue rate, in thousandths.
# 478.3633 302.3879 131.8317 67.308 20.10918

# Check on correspondence analysis results.
names(gobCA)
# Find: "rproj" "rcorr" "rcntr" "cproj" "ccorr" "ccntr"

nrow(gobCA$rproj); ncol(gobCA$rproj)
# Shows that there are 25 rows and 5 columns.

# Supplementary columns:
gobCASuppC <- caSuppCol(gobelets[,-3], gobelets[,3])

# Outputing projections of the supplementary column, viz. column 3:
gobCASuppC
# -0.1355843782  0.0078285577 -0.0005536928  0.0057148860 -0.025171509

# Plotting follows.

plot(gobCA$rproj[,1],gobCA$rproj[,2],type="n",xlab=
    "Factor 1 (47.8% inertia)",ylab="Factor 2 (30.2% inertia)")
text(gobCA$rproj[,1],gobCA$rproj[,2],dimnames(gobelets[,-3])[[1]])
lines( c(max(gobCA$rproj[,1]),min(gobCA$rproj[,1])), c(0,0) )
lines( c(0,0), c(max(gobCA$rproj[,2]),min(gobCA$rproj[,2])) )
title("25 gobelets: factors 1 and 2")

plot(gobCA$cproj[,1],gobCA$cproj[,2],type="n",xlab=
    "Factor 1 (47.8% inertia)",ylab="Factor 2 (30.2% inertia)")
lines( c(max(gobCA$cproj[,1]),min(gobCA$cproj[,1])), c(0,0) )
lines( c(0,0), c(max(gobCA$cproj[,2]),min(gobCA$cproj[,2])) )
text(gobCA$cproj[,1],gobCA$cproj[,2],dimnames(gobelets[,-3])[[2]])
text(gobCASuppC[1],gobCASuppC[2],"Ht",font=4)
title("6 gobelet attributes, and 1 supplementary")

# Hierarchical clustering now.  Followed by plots.
source("c:/hcluswtd.r")
```

```
kIJ <- gobelets[,-3]
fIJ <- kIJ/sum(kIJ)
fI       <- apply(fIJ, 1, sum)
fJ       <- apply(fIJ, 2, sum)
h1 <- hierclust(gobCA$rproj, fI)
h2 <- hierclust(gobCA$cproj, fJ)
plot(as.dendrogram(h1))
title("Hierarchical clustering of 25 goblets")
plot(as.dendrogram(h2))
title("Hierarchical clustering of 6 goblet attributes")

plot(c(gobCA$rproj[,1],gobCA$cproj[,1]),c(gobCA$rproj[,2],gobCA$cpro
     type="n",xlab="Factor 1 (47.8% inertia)",ylab="Factor 2 (30.2% i
text(gobCA$rproj[,1],gobCA$rproj[,2],dimnames(gobelets[,-3])[[1]])
text(gobCA$cproj[,1],gobCA$cproj[,2],dimnames(gobelets[,-3])[[2]],fo
text(gobCASuppC[1],gobCASuppC[2],"Ht",font=3)
lines( c(max(gobCA$rproj[,1]),min(gobCA$rproj[,1])), c(0,0) )
lines( c(0,0), c(max(gobCA$rproj[,2]),min(gobCA$rproj[,2])) )
title("25 gobelets, 6 attributes, 1 supplementary attribute")
```

4.2   Comparison of Prehistoric and Modern Groups of Canids

A 7×6 table, Table 4.2, cross-classifies 7 species or groups of canids (dogs) with a set of 6 measurements taken on the mandible (lower jaw) and the teeth [26, 55]. Skull measurements would provide a better basis for the taxonomy; but the measurements are adequate for the comparison of a prehistoric group with 6 modern groups. In fact the mandible and the teeth are better preserved than the cranium.

In the view of [26] there is little to be expected from the analysis of such data: the mutual proximity of the species and their possible overlapping become manifest only through the analysis of populations, not means.

From [55] there is also a 77×9 table cross-tabulating with 9 mandible measurements a set of 77 individuals, belonging to the prehistoric group or to 4 modern groups: this table (Table 4.3) will be analyzed.

The molar range consists of 3 teeth, except for the cuons, for which it consists of only two. The nine measurements are:

Mnd: length of mandible;

MndB: breadth of mandible below 1st molar;

Cndl: breadth of articular condyle;

TABLE 4.2

Mean mandible measurements (mm/10) for modern Thai dogs, golden jackals, wolves, cuons, dingos and prehistoric dogs (X1 = breadth of mandible, X2 = height of mandible below 1st molar, X3 = length of 1st molar, X4 = breadth of 1st molar, X5 = length from 1st to 3rd molars inclusive, X6 = length from 1st to 4th premolars inclusive).

Group	X1	X2	X3	X4	X5	X6
Modern dog	97	210	194	77	320	365
Golden jackal	81	167	183	70	303	329
Chinese wolf	135	273	268	106	419	481
Indian wolf	115	243	245	93	400	446
Cuon	107	235	214	85	288	376
Dingo	96	226	211	83	344	431
Prehistoric dog	103	221	191	81	323	350

```
Modern dog      _____8_____10_____12___
Chinese wolf    _____|        |                                   |
Prehist. dog    _____|                                    |
Indian wolf     _____9_____11_____
Golden jackal   _____|                              |
Dingo           _____|
Cuon            _____
```

FIGURE 4.4

Part of hierarchical clustering tree of canids.

TABLE 4.3

First three rows only of 77 × 9 data are shown: see web site,
www.correspondances.info, for remainder of the data.

	Mnd	MndB	Cndl	MndH	molL	molB	3mol	4prm	canB
Ta	1234	101	231	228	192	78	322	330	56
Tb	1274	96	190	219	192	78	322	404	58
Tc	1211	102	178	210	206	79	345	375	62

MndH: height of mandible below 1st molar;

molL: length of 1st molar;

molB: breadth of 1st molar;

3mol: length from 1st to 3rd molars inclusive (1st to 2nd for cuon);

4prm: length from 1st to 4th premolars inclusive;

canB: breadth of lower canine; all in mm/10.

The individuals (i.e., dogs) are divided up into a set Is of 9 groups. The 10 prehistoric specimens of Thai dogs are denoted by ? followed by a lower case letter from a to j: the gender of the individuals to which these mandibles belong is unknown.

For the four modern species, there is a group divided into 2 subgroups according to the gender. To each species we assign a letter: {T = Thai dog; C = Cuon; J = Jackal, H = Indian wolf}: a male individual is denoted by the letter for its species, in upper case, followed by a lower case letter to indicate its serial number; the contrary notation is adopted for the females. Thus Tc denotes the 3rd male individual of the species Thai dog, and cD denotes the 4th female individual of the species Cuon.

To the initial table, we adjoin rows, pertaining to the animal groups or subgroups, created by accumulating the rows describing the individuals: e.g., with the modern male dogs we associate a row §T obtained by accumulating the 8 individual rows; with the prehistoric dogs, we associate §?, obtained by accumulating the 10 individual rows; etc.

Correspondence analysis of this 77 × 9 table follows. The percentages of inertia explained by the major factors are, in decreasing order: 50%, 21%, 10%, 7%, 5%.

On axis 1 (see following listing, showing factor projections, contributions and correlations for the canid data), the length of the molar range and the length of the premolar range { 3mol, 4prm }, $F1 > 0$, is in contrast with the height of the mandible and with the breadth of the mandible and of the articular condyle: { Cndl, MndH, MndB }.

```
|SYMI|QLT WTS INR |  F1 CO2 CTR |  F2 CO2 CTR |  F3 CO2 CTR |  F4 CO2 CTR |
---------------------------------------------------------------------------
Below, supplementary elements
|supT| 846 104   7|   4 102   2|   8 350  12|   0   0   0|   9 394  41|
|supt| 679  99   5|  -1   3   0|  -6 244   6|   5 193  10|   6 240  18|
|supJ| 972 115  69|  38 875 121| -10  56  18|   6  19  13|  -6  23  22|
```

	QLT WTS INR	F 1 CO2 CTR	F 2 CO2 CTR	F 3 CO2 CTR	F 4 CO2 CTR
|supj|	871 110 76|	38 750 114|	-14 103 37|	6 16 12|	2 1 2|
|supC|	997 121 131|	-52 916 240|	-4 4 3|	-11 43 55|	-10 34 62|
|supc|	984 106 120|	-54 922 222|	2 1 1|	-10 33 39|	-9 28 48|
|supH|	662 129 21|	1 1 0|	-16 583 58|	6 78 16|	0 0 0|
|suph|	925 90 39|	32 844 66|	-7 36 7|	-7 42 16|	2 3 2|
|sup?|	974 126 87|	1 1 0|	42 931 383|	4 9 8|	8 33 40|

|SYMJ|	QLT WTS INR	F 1 CO2 CTR	F 2 CO2 CTR	F 3 CO2 CTR	F 4 CO2 CTR
| Mnd|	732 450 49|	-3 37 4|	-12 475 110|	-7 143 69|	5 76 52|
|MndB|	724 35 62|	-34 238 30|	42 353 105|	12 29 18|	-23 104 91|
|Cndl|	973 77 188|	-60 540 203|	-41 253 226|	35 180 334|	-1 0 1|
|MndH|	981 75 204|	-59 467 190|	58 453 439|	3 1 3|	21 60 172|
|molL|	685 71 46|	9 44 4|	1 1 0|	1 1 0|	-34 639 415|
|molB|	498 28 37|	-8 16 1|	31 255 45|	-3 2 1|	-29 225 118|
|3mol|	998 113 294|	78 861 507|	15 31 43|	27 101 293|	6 5 22|
|4prm|	687 130 76|	23 330 50|	-4 10 3|	-23 341 256|	3 6 6|
|canB|	519 21 44|	-26 123 11|	28 137 29|	-18 60 26|	-34 199 123|

As regards the species which figure here as supplementary elements, the male and female cuons $\{supC, supc\}$ contrast, on axis 1, with the jackals of both genders $\{supJ, supj\}$. At the center we find the Thai dogs, $\{supT, supt\}$, the male wolves, supH, and the prehistoric dogs, sup?. The female wolves, suph, are, on axis 1, close to the jackals; but the dispersion of the individual wolves prevents us from asserting anything as regards the species in general.

The cuons, whose molar range is reduced to two teeth, find themselves contrasted with the length of this molar range: this contrast may be regarded as an artifact related to the choice of the measurements [26].

On axis 2, sup? (prehistoric dogs), which has a near perfect correlation (COR = 931), is opposed to all the modern animal groups. More precisely, on axis 2, sup?, associated with MndH, is opposed to Cndl: it is possible that the articular condyle was not in a proper state of preservation and thus led to underestimation among the prehistoric specimens.

In the plot of the (1,2) plane, we see the three sets J (measurements: Mnd, MndB, Cndl, MndH, molL, molB, 3mol, 4prm, canB), Is (specific groups, put as supplementary: supT, supt, supJ, supj, supC, supc, sup?) and I (individuals: Ta, Tb, Tc, ... , ?h, ?i, ?j). The group of cuons (C) is in the half-plane $F1 < 0$; those of jackals (J) in $F1 > 0$. All the prehistoric mandibles are in $F2 > 0$. Thai dogs and Indian wolves are dispersed on a strip starting from a central area (set back towards $F2 < 0$) that extends into the quarter plane $F1 > 0, F2 < 0$. There is, for each of these species, a distance separating the center of the males and the center of the females, without clear discrimination between the genders which are present throughout the strip. This conclusion will be confirmed by the subsequent analyses.

The classification of the 77 mandibles is now looked at. A cluster analysis was carried out without taking into account the species or the gender, but only the shape, described by the 9 measurements. If specific groups could be

77 mandibles, 9 supplementary (groups), crossed by 9 attributes

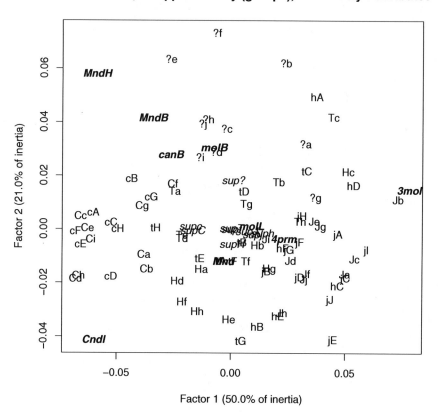

FIGURE 4.5

Factor plane (1,2) for canid data.

identified in the clusters, it would prove not only that they can be discriminated, but also that they are determined, without any a priori information, solely by the inductive study of the mandibles.

The partition resulting from classification of canid data follows.

```
Number of clusters: 12
Cluster contents follow.
```

```
Cluster i106: Hg hE
Cluster i126: tC xa Tc hA
Cluster i127: Jb Je Jc jA
Cluster i132: cE cF Cc cA cD Ca cH Cd Ch Ce Ci
Cluster i134: Tb Tg hF tB Hb
Cluster i135: Ta tH Hd Hf Td tA
Cluster i137: jG Jd jB tD jH Jh Jf Jj jD Tf Ji jF
Cluster i138: Hh Te tE tF Ha He
Cluster i139: Cb cB Cg cG Cf cC
Cluster i140: ?b ?f ?e ?d ?c ?h ?i ?j
Cluster i141: Ja hC hD Jg jI xg Th Hc
Cluster i142: tG hB jC jE jJ
```

```
Subsequent agglomerations.
```

```
i106 and i137    -->    i143
i135 and i138    -->    i144
i141 and i127    -->    i145
i132 and i139    -->    i146
i134 and i143    -->    i147
i126 and i145    -->    i148
i142 and i147    -->    i149
i144 and i149    -->    i150
i140 and i148    -->    i151
i150 and i151    -->    i152
i146 and i152    -->    i153
```

The enumeration of the individuals in the various clusters shows first of all that the set of 17 cuons, both male and female, constitute class i46 which separates itself from the rest of the individuals at the highest level of the hierarchy. But we have already noted that the cuons distinguish themselves a priori by the shape of their molar range.

Within class i152, three subdivisions { i150, i140, i148 } can be distinguished, of which only one can be interpreted clearly: i140 contains 8 out of 10 prehistoric mandibles. At the lowest level of the hierarchy that we have chosen to consider, certain classes are very nearly confined to a single species: i137 comprises 10 jackals, with 2 Thai dogs; and i127 is constituted by 4 jackals; but no class is identified with a species.

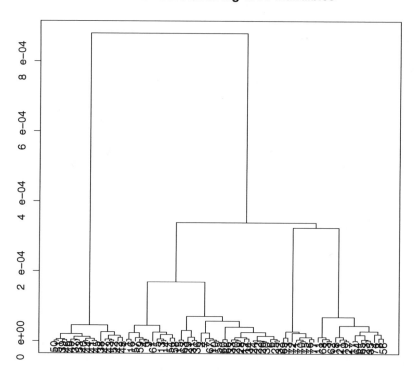

FIGURE 4.6

Dendrogram derived from canid data.

Discriminant analysis (supervised classification) serves to validate clustering results and to extrapolate to unseen cases. In the Euclidean space of the profiles, referred to the 8 axes obtained from correspondence analysis, the 77 mandibles are projected as principal elements, and the 9 animal groups as supplementary. We then allocate each individual to the group which is closest to it.

```
Allocation of individuals to centers.
Number of factors used = 8
Allocation of the 77 individuals i to the 9 centers
(4x2 present-day + 1 fossil)

   Tb   Te   Tg   tD1                                 --> supT
   Ta   Td   Tf   tA   tB   tE   tF   tG   tF  [Ji]    --> supt
  [tC]  Jb   Jc   Jd   Je   Jf   Jh   jA   jI          --> supJ
   Jj   jB   jC   jD   jE   jF   jG   jH   jJ          --> supt
   Ca   Cb   Cd   Cf   Ch   Ci   cD                    --> supC
   Cc   Ce   Cg   cA   cB   cC   cE   cF   cG   cH      --> supC
   Ha   Hb   Hd   He   Hf   Hg   Hh   hB   hE          --> supH
  [Th   Ja   Jg]  Hc   hC   hD   hF  [?g]              --> suph
  [Tc   hA]  ?a   ?b   ?c   ?d   ?e   ?f   ?h   ?g  ?h  --> sup?
```

We display the list of the individuals allocated to each center; indicating within square brackets the rare disagreements (in all 8 out of 77) relative to the species. As regards the gender, the agreement is not quite so good. However in most of the species the division is not random: thus, 6 males and 1 female are allocated to supC; and 7 females and 3 males to supc.

```
       Table of allocations: insupx to supy.
```

	insupT	insupt	insupJ	insupj	insupC	insupc	insupH	insuph	insup?
--> supT	3	1	0	0	0	0	0	0	0
--> supt	3	6	1	0	0	0	0	0	0
--> supJ	0	1	6	2	0	0	0	0	0
--> supj	0	0	1	8	0	0	0	0	0
--> supC	0	0	0	0	6	1	0	0	0
--> supc	0	0	0	0	3	7	0	0	0
--> supH	0	0	0	0	0	0	7	2	0
--> suph	1	0	2	0	0	0	1	3	1
--> sup?	1	0	0	0	0	0	0	1	9

The table of allocations, constructed from the list, confirms the stability of the animal groups, represented by the approximate square blocks of non-zero values along and close to the diagonal.

The success of the discriminant analysis which we have just described encourages us to repeat the discrimination by reserving a test sample. We know that, for this, the set I of the individuals is divided between Ii, the initial sample, and It, the test sample. The center of gravity of each species (or, here, subspecies) is calculated from the individuals of Ii.

More precisely, we analyze the principal table crossing Is with J; in which the row pertaining to each species §s is the accumulated row of its individuals belonging to Ii; and each individual, from Ii or from Is, is allocated to the center of $I\S$ which is the closest to it.

The percentages of inertia for the main factors are found to be (in decreasing order): 72%, 18%, 4%, 3%.

Confusion matrix for the test sample.

	insupT	insupt	insupJ	insupj	insupC	insupc	insupH	insuph	insup?
--> supT	0	0	0	0	1	0	0	0	0
--> supt	1	2	0	0	0	0	2	0	0
--> supJ	0	0	2	3	0	0	0	0	0
--> supj	0	0	0	1	0	0	0	0	0
--> supC	0	0	0	0	1	1	0	0	0
--> supc	0	0	0	0	1	1	0	0	0
--> supH	0	1	0	0	0	0	1	2	0
--> suph	1	0	1	0	0	0	0	0	0
--> sup?	0	0	0	0	0	0	0	0	3

The centers being determined from Ii, we expect the allocation to be more accurate for this set than for It; but, additionally, it is the allocation of It which alone can show how reliable our system of centers is for allocating any individuals whatever, external to the data of the present study. Here the test sample comprises 25 individuals (those whose order – from 1 to 77 – is a multiple of 3: i.e., 3, 6, ... , 75).

Outside of the test sample, i.e., for the 52 individuals whose order is not a multiple of 3, there are, as regards the species, only 8 errors of allocation.

For the test sample, the number of correct allocations as regards the species is 19 out of 25, or more than three quarters, which cannot be due to chance. On the other hand, out of 16 correct assignments as regards the species and involving the individuals of the non-prehistoric era, whose gender is known, 8 are erroneous as regards the gender: the gender is therefore in no way recognized in the test sample.

Conclusion: Discrimination of the Species of Canids

Multidimensional analysis has shown that the group of cuons, whose molar range measured consisted only of two teeth, cannot usefully be considered with the other species: we must therefore seek the conclusions of the study by confining ourselves to the 60 individuals of the other 7 groups.

We use first the cluster analysis results, as follows, carried out through analysis of the 60×9 table, as we did earlier for the case of the 77×9 table.

A partition defined from the 60 x 9 canid data set.

Number of clusters: 10
Cluster contents follow.

Cluster i81: Hg hE
Cluster i97: tC ?a Tc hA
Cluster i98: Jb Je Jc jA

```
Cluster i99:  Tf Ji jF jG Jd jB tD jH
Cluster i102: Jh Jf Jj jD jC jE jJ
Cluster i104: Tb Tg hF tB Hb
Cluster i105: Ta tH Hd Hf Td tA
Cluster i108: ?b ?f ?e ?d ?c ?h ?g ?h
Cluster i109: Ja hC hD Jg jI ?g Th Hc
Cluster i110: tG hB Hh Te tE tF Ha He
```

Subsequent agglomerations.

```
 i98 and i102    -->   i111
 i81 and i104    -->   i112
i110 and i112    -->   i113
 i99 and i111    -->   i114
 i97 and i109    -->   i115
i105 and i113    -->   i116
i114 and i115    -->   i117
i108 and i116    -->   i118
i117 and i118    -->   i119
```

For the species considered here, the cluster analysis does not very much differ from the one made with the cuons present. The prehistoric mandibles are well recognized in class 108; and the jackals in class 114. No separation appears between Thai dogs and wolves, which are intimately mixed in class 116 as in class 115.

We then carry out a discriminant analysis with a test sample. As before, the principal table cross-classifies a set J of measurements with the groups, numbering 7 here, whose profiles are defined by accumulating the individuals of the group in Ii, while reserving It as test sample. Again as before, we have, in the test sample, a high proportion of recognition of the species: the proportion of errors is only 5 out of 20.

To be precise, the errors are as follows: (Tc $\longrightarrow$ X suph), (tG $\longrightarrow$ X supH), (Jb $\longrightarrow$ X suph), (Ha $\longrightarrow$ X supt) and (Hd $\longrightarrow$ X supt). Leaving aside the male jackal Jb allocated to the group of the female wolves, §h, these errors concern the confusion between dogs and wolves.

Since it has appeared, above, that the gender cannot be recognized from the shape, we have attempted to do a discriminant analysis with the test sample, not with the 7 centers of the groups, but with four centers of the species, viz. of Tt, Jj, Hh and ?. We were surprised to find that the results were clearly less satisfactory than with the seven centers. This experiment conveys an important conclusion, which we shall interpret a posteriori. In the space of the shapes, as we have seen as soon as we did the correspondence analysis at the start of our analysis of canids, a species is not necessarily concentrated around a point: it can be dispersed over a strip. The discrimination will be better if, for such a species, we have two centers which account for its express spatial extension.

There remains the confusion between the dogs and the wolves. It turns out that this confusion, occurring when the shape is considered, is resolved when we consider the size. In the Thai dogs of both genders, the length Mnd of the mandible exceeds 130.1 mm only in one case; in the Indian wolves, on the contrary, Mnd exceeds 140.9 mm, except in one case.

It is known that as compared to the skull, the mandible is much more developed in the wolf than in the dog: thus, the size difference that one notices here in a sample of mandibles arises from a shape difference, and this is a criterion that is much more satisfactory, considering the great variability of size of the domestic dogs. What is implied here is that this much more satisfactory criterion has not been made available to us. If we had had more complete measurements, i.e., including the skull measurements, correspondence analysis, which analyzes shapes, would have highlighted the difference between the dog and the wolf; whereas, if we considered the mandible measurements alone, as here, we can never conclude that a mandible belongs to a wolf because it is large, given that some domestic dogs also have large mandibles.

It therefore seems that the problem of discrimination posed has been resolved. But recall that we have excluded the cuons, with their molar range of 2 teeth; and that the opposition to Cndl, shown by prehistoric dogs, may be due to a deterioration over time of the condyles.

Projections, correlations and contributions.

|SIMI|	QLT	WTS	INR	F 1	CO2	CTR	F 2	CO2	CTR	F 3	CO2	CTR
|supT|	837	134	41	14	787	49	-3	48	8	1	3	2
|supt|	922	128	56	5	83	7	-15	655	142	8	183	167
|supJ|	935	148	134	-23	735	151	12	199	104	-1	1	2
|supj|	981	143	168	-27	769	198	11	117	77	10	95	261
|supH|	971	167	129	-3	11	2	-24	938	475	-4	22	47
|suph|	946	117	72	-15	416	46	7	92	26	-15	438	521
|sup?|	997	163	400	42	890	547	15	107	167	0	0	0

|SIMJ|	QLT	WTS	INR	F 1	CO2	CTR	F 2	CO2	CTR	F 3	CO2	CTR
	Mnd|	968	449	54	-5	217	18	-7	475	101	-5	275	246
|MndB|	962	34	84	44	945	121	4	7	2	4	9	12
|Cndl|	981	75	129	-9	57	11	-33	749	379	16	175	372
|MndH|	996	73	399	66	978	599	7	10	16	6	8	54
|molL|	935	71	53	-23	830	68	8	104	22	1	1	1
|molB|	780	28	42	22	380	25	22	384	64	-5	16	11
|3mol|	980	118	110	-14	261	44	23	717	308	1	2	3
|4prm|	908	131	61	-14	484	45	11	291	70	7	134	135
|canB|	970	21	67	42	673	69	-20	145	38	-20	152	166

Lastly, for purposes of comparison with the cluster analysis given at the beginning of section 4.2, we give below the results of the cluster analysis as well as the correspondence analysis of the 7×9 table in which each row is the accumulation of all the individuals of a group. The percentages of inertia explained by the most important factors are: 65%, 25.5%, 6%, 2%.

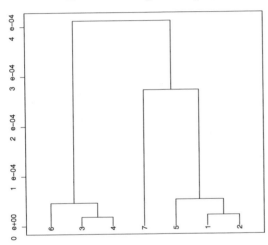

Hierarchical clustering of 7 canid groups

FIGURE 4.7

Hierarchical clustering of species (7×9 table). Canid groups: 6, 3, 4, 7, 5, 1, 2 which represent, respectively, supH, subJ, supj, sup?, supH, supT, supt.

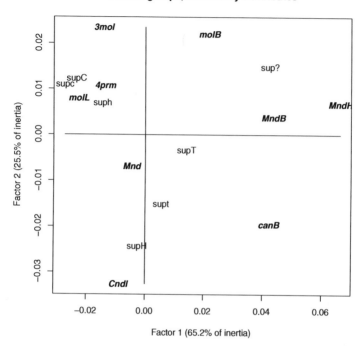

FIGURE 4.8
Principal factor plane (1,2) of the 7 × 9 data set.

In the cluster analysis of the species (or subspecies), the prehistoric group, ?, is isolated. The two genders of the jackals {$supJ, supj$} are agglomerated, as are those of the dogs {$supT, supt$}, although one can see, in the (1,2) plane, a distance separating them; but supH agglomerates with the dogs, suph with the jackals: thus, male wolves and female wolves go into different clusters: recall that we have already remarked on the dispersion of this species.

The variables { MndH, Cndl, 3mol } each present in one of the three major clusters of variables, viz. { j15, j12, j14 }, bring major contributions to the cluster analysis of the species.

4.2.1   Software Used

R is used for the analysis of the mandible data, on a Windows platform.

```
# Input our data and the programs to be used.
x <- read.table("c:/mandible77s.dat")
```

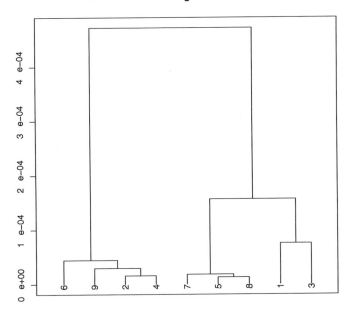

Hierarchical clustering of 9 canid attributes

FIGURE 4.9

Hierarchical clustering of the measurements (7×9 data set). Attributes: 6, 9, 2, 4, 7, 5, 8, 1, 3 = molB, canB, MndB, MndH, 3mol, molL, 4prm, Mnd, Cndl.

```
nrow(x); ncol(x)
# Checking on numbers of rows and columns furnishes: 86 x 9.
# We used "mol3" and "prm4" since our Java input
#    syntax required this.
# Switch back to original numeric-leading character:
dimnames(x)[[2]][7] <- "3mol"
dimnames(x)[[2]][8] <- "4prm"
# Correspondence factor analysis follows.
xca <- ca(x[1:77,])
# Supplementary rows analysis follows.
xcar <- caSuppRow(x[1:77,], x[78:86,])

# Plots follow.
plot(c(xca$rproj[,1],xca$cproj[,1]),
     c(xca$rproj[,2],xca$cproj[,2]),type="n",
   xlab="Factor 1 (50.0% of inertia)",
   ylab="Factor 2 (21.0% of inertia)")
text(xca$rproj[,1],xca$rproj[,2],dimnames(x)[[1]])
text(xca$cproj[,1],xca$cproj[,2],dimnames(x)[[2]],font=4)
text(xcar[,1],xcar[,2],dimnames(xcar)[[1]],font=3)
title
 ("77 mandibles, 9 supplementary (groups), crossed by 9 attributes")

# Hierarchical clustering follows.
source("c:/hcluswtd.r")
kIJ      <- x[1:77,]
fIJ      <- kIJ/sum(kIJ)
fI       <- apply(fIJ, 1, sum)
fJ       <- apply(fIJ, 2, sum)
h1 <- hierclust(xca$rproj, fI)
h2 <- hierclust(xca$cproj, fJ)
plot(as.dendrogram(h1))
title("Hierarchical clustering of 77 mandibles")
```

We use our Java program on the same data, now. The input data file, mandible77s.dat, starts as follows.

Mandibles 77 canines in 5 types, crossed by 9 characteristics								
86 9 SUPR 8 12	6 Mnd	MndB	Cndl	MndH	molL	molB	mol3
					prm4	canB	SUPP
Ta	1234	101	231	228	192	78	322	
						330	56 1	
Tb	1274	96	190	219	192	78	322	
						404	58 1	

The data set is cut following line 4. (Note the wrap-around used above, for display purposes.) The full data set is available on the book's web site, www.correspondances.info. In the above we requested output for a 12-cluster solution, which provides the following.

```
Number of clusters: 12
Cluster contents follow.

Cluster i106: Hg hE
Cluster i126: tC xa Tc hA
Cluster i127: Jb Je Jc jA
Cluster i132: cE cF Cc cA cD Ca cH Cd Ch Ce Ci
Cluster i134: Tb Tg hF tB Hb
Cluster i135: Ta tH Hd Hf Td tA
Cluster i137: jG Jd jB tD jH Jh Jf Jj jD Tf Ji jF
Cluster i138: Hh Te tE tF Ha He
Cluster i139: Cb cB Cg cG Cf cC
Cluster i140: ?b ?f ?e ?d ?c ?h ?i ?j
Cluster i141: Ja hC hD Jg jI xg Th Hc
Cluster i142: tG hB jC jE jJ

Subsequent agglomerations.

i106 and i137    -->   i143
i135 and i138    -->   i144
i141 and i127    -->   i145
i132 and i139    -->   i146
i134 and i143    -->   i147
i126 and i145    -->   i148
i142 and i147    -->   i149
i144 and i149    -->   i150
i140 and i148    -->   i151
i150 and i151    -->   i152
i146 and i152    -->   i153
```

The output for a 10-cluster solution is as follows.

```
Number of clusters: 10
Cluster contents follow.

Cluster i81: Hg hE
Cluster i97: tC ?a Tc hA
Cluster i98: Jb Je Jc jA
```

```
Cluster i99: Tf Ji jF jG Jd jB tD jH
Cluster i102: Jh Jf Jj jD jC jE jJ
Cluster i104: Tb Tg hF tB Hb
Cluster i105: Ta tH Hd Hf Td tA
Cluster i108: ?b ?f ?e ?d ?c ?h ?g ?h
Cluster i109: Ja hC hD Jg jI ?g Th Hc
Cluster i110: tG hB Hh Te tE tF Ha He
```

Subsequent agglomerations.

```
 i98 and i102     -->    i111
 i81 and i104     -->    i112
i110 and i112     -->    i113
 i99 and i111     -->    i114
 i97 and i109     -->    i115
i105 and i113     -->    i116
i114 and i115     -->    i117
i108 and i116     -->    i118
i117 and i118     -->    i119
```

Correspondence analysis in R now follows. Given that we want output plots, doing this in R is best.

```
y <- rbind(x[78:79,],x[82:86,])
# Some checks.
nrow(y); ncol(y); dimnames(y)[[1]]; dimnames(y)[[2]]
# Results: 7   9
# "supT" "supt" "supC" "supc" "supH" "suph" "sup?"
# "Mnd"  "MndB" "Cndl" "MndH" "molL" "molB" "3mol" "4prm" "canB"
fIJ <- y/sum(y)
fI      <- apply(fIJ, 1, sum)
fJ      <- apply(fIJ, 2, sum)
yca <- ca(y)
# Eigenvalues follow (trivial first eigenvalue removed).
# 0.0008980387 0.0003414525 4.376261e-05 1.837993e-05
#              6.117846e-06 1.335595e-06 0
# Eigenvalue rate, in thousandths.
# 686.0037 260.8325 33.42987 14.04026 4.673368 1.020249 0 0
h1 <- hierclust(yca$rproj[,1:6], fI)
plot(as.dendrogram(h1))

z <- y
z[3,] <- apply(x[17:26,],2,sum)
z[4,] <- apply(x[27:36,],2,sum)
```

```
zca <- ca(z)
# Eigenvalue rate:
# 652.0931 255.3553 60.84598 21.05068 8.254496 2.400418 0 0

# Some plots next.

plot(c(zca$rproj[,1],zca$cproj[,1]),
     c(zca$rproj[,2],zca$cproj[,2]),type="n",
   xlab="Factor 1 (65.2% of inertia)",
   ylab="Factor 2 (25.5% of inertia)")
text(zca$rproj[,1],zca$rproj[,2],dimnames(z)[[1]])
text(zca$cproj[,1],zca$cproj[,2],dimnames(z)[[2]],font=4)
lines( c(max(zca$cproj[,1]),min(zca$rproj[,1])), c(0,0) )
lines( c(0,0), c(max(zca$cproj[,2]),min(zca$cproj[,2])) )
title("7 canid groups, crossed by 9 attributes")

fIJ <- z/sum(z)
fI  <- apply(fIJ, 1, sum)
h1  <- hierclust(zca$rproj[,1:6],fI)
plot(as.dendrogram(h1))
title("Hierarchical clustering of 7 canid groups")

# Canid groups:
# 6, 3, 4, 7, 5, 1, 2 =  supH, subJ, supj, sup?, supH, supT, supt

fJ <- apply(fIJ, 2, sum)
h2 <- hierclust(zca$cproj[,1:6],fJ)
plot(as.dendrogram(h2))
title("Hierarchical clustering of 9 canid attributes")

# Attributes:
# 6, 9, 2, 4, 7, 5, 8, 1, 3 = molB, canB, MndB, MndH,
#      3mol, molL, 4prm, Mnd, Cndl
```

4.3 Craniometric Data from Ancient Egyptian Tombs

The data [26, 55] concern human skulls, of subjects presumed to be males, found in Egypt, in ancient tombs in the region of Thebes. The set I of the skulls is divided according to the period into a set Is of 5 series of 30:

1–30: to -4000

TABLE 4.4

Ancient Egyptian skulls data
set. First three rows shown:
remainder of data on web
site.

skull	Br	Hh	Ba	Ns
1	131	138	89	49
2	125	131	92	48
3	131	132	99	50

31–60: to −3300
61–90: to −1850
91–120: to −200
121–150: to 150

The first series goes back to the early predynastic period, that is about 4000 BC. Series 2 is of the late predynastic period, that is 3300 BC. Series 3 is placed at about 1850 BC, under the twelfth and the thirteenth dynasties. Series 4, 200 BC, belongs to the Ptolemaic period. The last, 5, belongs to the Roman period.

On each skull, a set J of four measurements have been made. As early as 1973, much more precise craniometric data were analyzed using correspondence analysis, and a highly pertinent discussion on the choice of the measurements may be found in the research work of Louis Bellier ([15], Part C, Sections 5 and 6). Here we have $J = \{$ Br, Hh, Ba, Ns $\}$, respectively maximum breadth, basibregmatic height, basialveolar length and nasal height.

We have then a table of measurements $I \times J$, $(5 \times 30) \times 4$, to which we add a set Is of 5 rows, each of which is the accumulated row of the descriptive rows pertaining to the 30 individuals of a series. The percentages of inertia are found to be: 48%, 27% and 24%.

Next, we examine variation of the shape of the skulls over time. The very structure of the sample I attests to the fact that the data have been collected for detecting a possible variation over time in the shape of the skulls. Now this variation is very pronounced: it appears at once in the analysis of the principal table $I \times J$, from the location of the set Is of supplementary elements.

The opposition between predynastic, 1–30 and 31–60, and postdynastic, 90–120, 121–150, can be read accurately on axis 1, with the CO2 (relative contributions, i.e., squared correlation coefficients, in thousandths) higher than 929. The series belonging to the middle empire, 61–90, emerges on axis 3; so that the image of the (1,3) plane obtained from $I \times J$ does not differ from what would be obtained from the (1,3) plane of $Is \times J$. In the course of thousands of years, Ba tends to get smaller as compared to the breadth Br.

The differences between the series are highly significant, as one can verify by projecting a cloud Iha of 200 fictitious (simulated) samples, each represented by the accumulation of 30 individuals drawn at random.

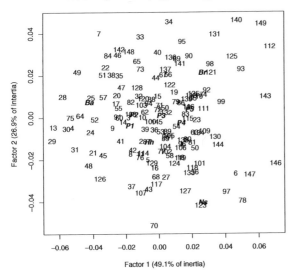

FIGURE 4.10
Correspondence analysis principal factor plane (1,2).

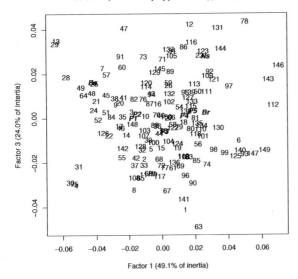

FIGURE 4.11
Correspondence analysis principal factor plane (1,3).

Projections, correlations and contributions for
Egyptian skull data.

```
-----------------------------------------------------------------
|SYMI| QLT WTS INR|  F 1 CO2 CTR|   F 2 CO2 CTR|   F 3 CO2 CTR|
-----------------------------------------------------------------
below, supplementary elements
|   1|1000 200  44|  -18 957  86|  -4  43   7|    0   0   0|
|   2|1000 200  32|  -15 976  64|   1   8   1|   -2  15   2|
|   3|1000 200   5|    1  39   0|   1  22   0|    6 939  20|
|   4|1000 200  25|   13 962  49|  -2  33   3|   -1   5   1|
|   5|1000 199  49|   19 930  93|   4  42   8|   -3  27   6|


-----------------------------------------------------------------
|SYMJ| QLT WTS INR|  F 1 CO2 CTR|   F 2 CO2 CTR|   F 3 CO2 CTR|
-----------------------------------------------------------------
|  Br|1000 324 249|   26 600 305|   21 395 366|   -2   5   6|
|  Hh|1000 320 170|   -6  45  16|  -12 168 106|   25 787 558|
|  Ba|1000 233 326|  -42 847 564|    7  27  32|  -16 126 171|
|  Ns|1000 123 254|   26 224 116|  -40 525 495|  -28 251 266|
-----------------------------------------------------------------
```

We now follow with an attempt to recognize the period of a skull from its shape. We cannot expect the shape differences between the series to be of such amplitude that the period to which a skull belongs could be known for certain from four measurements. It is however worth the trouble to consider the results of a discriminant analysis based on the correspondence analysis which has just been described.

Each element i of I is allocated to the series is to whose center it is closest (in the profile space, with the χ^2 distance; i.e., in the Euclidean space in which F1, F2, F3 constitute an orthonormal system of coordinates).

Discriminant analysis based on correspondence analysis.

	1--30	31--60	61--90	91--120	121--150
c1	12	9	5	3	2
c2	7	8	3	4	4
c3	5	7	13	4	4
c4	4	5	3	7	8
c5	2	1	6	12	12

Results obtained when 150 Egyptian skulls are allocated to the groups for which they have the minimum Mahalanobis distance.

	1--30	31--60	61--90	91--120	121--150
c1	12	10	4	3	2
c2	8	8	4	3	4
c3	4	5	15	7	4
c4	4	4	2	5	9

| c5 | 2 | 3 | 5 | 12 | 11 |

In the cumulative table, we see, e.g., that 7 individuals of series 1 are attached to the center c2 of series 2; 12 individuals of series 5 are correctly allocated to the center c5, and so on.

From the Mahalanobis distance, we obtain an allocation matrix, which is not very much different. The total of the diagonal terms (accurate allocations) is 51, as against 52 above. Serious errors, i.e., allocation of predynastic skull to postdynastic period and vice versa (interchanges between series 1, 2 on the one hand, and series 4, 5 on the other), are 25 in number, as against 25 also above.

These numerical differences are insignificant. What is more important is that all the interesting results appear in an orderly fashion in the space of the profiles with the χ^2 distance and referred to the axes determined by correspondence analysis.

4.3.1 Software Used

For the Egyptian skulls data, we show here the R programs used.

```
# Labels:
# a1   -   a30
# b31  -   b60
# c61  -   c90
# d91  -   d120
# e121 -   e150

skulls <- read.table("c:/skulls.txt")

# Check on numbers of rows, columns.
nrow(skulls); ncol(skulls)
# Find: 155, 4
# Set labels:
dimnames(skulls)[[1]] <-
     c(as.character(1:150),"P1","P2","P3","P4","P5")

# Correspondence analysis.
skca <- ca(skulls[1:150,])
# Eigenvalues follow (trivial first eigenvalue removed).
# 0.0007341408 0.0004027129 0.0003595096
# Eigenvalue rate, in thousandths.
# 490.6167 269.1278 240.2555

# Plots follow.
plot(c(skca$rproj[,1],skca$cproj[,1]),
```

```
      c(skca$rproj[,2],skca$cproj[,2]),
      type="n",xlab="Factor 1 (49.1% of inertia)",
      ylab="Factor 2 (26.9% of inertia)")
text(skca$rproj[,1],skca$rproj[,2],
    dimnames(skulls[1:150,])[[1]])
text(skca$cproj[,1],skca$cproj[,2],
    dimnames(skulls[1:150,])[[2]],font=4)

# Supplementary rows.
skcar <- caSuppRow(skulls[1:150,], skulls[151:155,])
text(skcar[,1],skcar[,2],dimnames(skulls[151:155,])[[1]],font=4)
title("150 skulls, 5 centers (supplementary), 4 attributes")

plot(c(skca$rproj[,1],skca$cproj[,1]),
      c(skca$rproj[,3],skca$cproj[,3]),
    type="n",xlab="Factor 1 (49.1% of inertia)",
    ylab="Factor 3 (24.0% of inertia)")
text(skca$rproj[,1],skca$rproj[,3],
    dimnames(skulls[1:150,])[[1]])
text(skca$cproj[,1],skca$cproj[,3],
    dimnames(skulls[1:150,])[[2]],font=4)
text(skcar[,1],skcar[,3],
    dimnames(skulls[151:155,])[[1]],font=4)
title("150 skulls, 5 centers (supplementary), 4 attributes")
```

4.4 Time-Varying Data Analysis: Examples from Economics

4.4.1 Imports and Exports of Phosphates

In [14, 54] analysis is carried out of import and export trade flows in phosphates. The former work is oriented towards the analysis of the trade data, whereas the latter is an extensive introduction to correspondence analysis based on this data.

Most phosphates, P_2O_5, are used in agricultural fertilizers, although some percentage finds other uses such as in animal foodstuffs, detergents, phosphor production (for matches), emulsifiers in the textile sector and others.

The original data used was comprised of the following, with a statement of the objective of the analysis.

1. The set T of 8 years, from 1973 to 1980.

2. The set JX of 9 main phosphate exporting countries.

3. The set $JX2$ of 22 second order exporting countries. These were used as supplementary elements.

4. The value $k(jx, t)$ is the volume of exports from country jx during year t.

5. The profile over time of countries is plotted in the correspondence or factor analysis display.

6. The 8 years are also plotted, showing the global evolution.

A similar analysis is carried out for the import countries. There are 22 main import countries. The objectives of the analysis of the imports is analogous to the analysis of the exports.

Analysis follows in [54] of the quaternary data table crossing: J, the set of principal countries; T, the set of successive years; M, the set of four movements: production, consumption, exportation and importation; and F, the set of forms of phosphate, namely: mineral, acid and fertilizer. Therefore the value $k(j, m, t, f)$ is the movement m (e.g., production) of phosphate in the form f relating to the country j in the year t. With more certainty of the data contained in $J \times M \times T$, a number of analyses of marginals were carried out and discussed in detail in [54]: $J \times M$; $(J \cdot T) \times M$ and $J \times (M \cdot T)$.

In [14], the ternary data table crossing the following was used.

- I, the set of 14 importing countries: Belgium (iBL), Canada (iCA), France (iFR), Germany (iDL), Italy (iIT), Japan (iJP), The Netherlands (iNL), Spain (iSP), United Kingdom (iUK), India (iIN), Brazil (iBR), Poland (iPL), Romania (iRM) and other eastern European countries (iEE).

- J, the set of 8 exporting countries: Belgium (eBL), USA (eUS), Jordan (eJR), Marocco (eMR), Senegal (eSN), Togo (eTG), Tunisia (eTN) and the USSR (eCC).

- T, the 8 years from 1973 to 1980.

The value $k(i, j, y)$ in the ternary data table, k_{IJT}, is the amount of phosphate, in thousands of tons, exported from country j to importing country i in year t.

Let us look at the data table based on imports and exports, $I \times J$, or table k_{IJ}, which accumulates the data over time T. This data crosses 14 countries by 8 countries. The dataset is available on the book's web site.

The factor analysis output listing follows. The number of factors requested is 7: for display convenience we show 4. The number of clusters requested for I, importing countries, is 4, and the number of clusters requested for J, exporting countries, is 8.

```
Commerce mondial des phosphates
14 8 SUPNO 7 4 8  eBL    eUS   eJR    eMR     eSN    eTG    eTN  eCC
iBL     0 1305    0   3573    25    500    110  293
[Data cut...]


phosphates.dat
Commerce mondial des phosphates
trace  :  1.1124
Lambda, rates and accumulative values are scaled by 10^-4
rank   :    1      2      3      4      5      6      7
lambda :  4505   3161   1524    819    575    335    206
rates  :  4050   2842   1370    736    517    301    185
cumul  :  4050   6891   8261   8997   9514   9815  10000
```

IDNI	QLT	WTS	INR	F#1	CO2	CTR	F#2	CO2	CTR	F#3	CO2	CTR	F#4	CO2	CTR
iBL	1000	60	22	-20	1	0	405	390	31	359	306	50	294	206	63
iCA	1000	86	139	-786	342	117	-1038	595	291	-130	9	9	266	39	74
iFR	1000	150	102	-208	57	14	649	556	200	-498	328	245	-153	31	43
iDL	1000	90	50	60	6	1	-242	94	17	-516	427	157	108	19	13
iIT	1000	59	15	-226	180	7	209	155	8	318	357	39	52	9	2
iJP	1000	70	37	-514	450	41	-473	381	50	223	84	23	-190	62	31
iNL	1000	58	51	-242	59	7	523	279	50	-444	200	74	6	0	0
iSP	1000	57	77	9	0	0	767	389	105	823	449	252	436	126	131
iUK	1000	49	56	-156	19	3	716	410	80	192	29	12	-47	2	1
iIN	1000	52	63	-398	117	18	-411	124	28	472	164	76	-891	585	502
iBR	1000	68	44	-569	444	49	-554	422	66	53	4	1	270	100	60
iPL	1000	82	29	559	802	57	95	23	2	146	55	11	4	0	0
iRM	1000	43	30	643	525	39	-179	41	4	380	184	40	-381	184	76
iEE	1000	78	284	1933	922	647	-523	67	67	-129	4	8	56	1	3

IDNJ	QLT	WTS	INR	F#1	CO2	CTR	F#2	CO2	CTR	F#3	CO2	CTR	F#4	CO2	CTR
eBL	1000	32	67	-146	9	2	440	84	20	-1151	575	281	-47	1	1
eUS	1000	355	201	-528	443	220	-585	543	385	-50	4	6	76	9	25
eJR	1000	27	80	1	0	0	-466	67	19	785	190	111	-1519	713	773
eMR	1000	337	122	45	5	2	492	602	259	371	342	304	134	45	74
eSN	1000	37	64	-299	46	7	768	306	68	-432	97	45	-460	109	94
eTG	1000	61	95	-156	14	3	822	387	129	-692	275	190	-136	11	14
eTN	1000	42	32	101	12	1	450	239	27	-407	195	46	-184	40	17
eCC	1000	108	339	1785	913	765	-523	78	94	-155	7	17	34	0	2

Relative to I, importing countries, the strong correlation (0.922) of the first factor with the diverse eastern European countries (iEE) means that this factor is strongly influenced by iEE. Relative to J, exporting countries, the strong correlation (0.913) of the first factor with the USSR (eCC) means that this factor is strongly influenced, overall, by exports from the latter to the eastern European importing country set.

If the correlations (CO2) tell us essentially how the factors, or axes, are determined, the contributions (CTR) tell us what factors are fundamentally determined by what observations or variables. So, with correlations, we think from the factors to the observations or variables; whereas with contributions we think from the observations or variables in the direction of the factors. In the analysis results above we see that the contributions to factor 1 are high on the part of iEE and eCC. So in this case the correlations and the contributions of factor 1 all point in the same direction. This is by no means always so: often enough the correlations and the contributions have a somewhat different tale to tell.

We will not look at the graphical displays, nor the hierarchical clustering, here. Using the appropriate data table, this can be quickly done with the Java programs. Or, using another appropriately set up data table, this can also be done in R.

We will just retain, for our use here, the description of the clusters.

```
Number of clusters: 4
Cluster contents follow.

Cluster i14: iEE
Cluster i20: iFR iNL
Cluster i22: iIN iCA iJP iBR
Cluster i24: iUK iSP iBL iIT iDL iPL iRM
```

From the following output we see (three leftmost columns) that clusters 20 and 24 merge into the next cluster up, cluster 25; then that new cluster, 25, is merged with cluster 22 in the next cluster up, cluster 26; and finally that new cluster, 26, is merged with cluster 14, in the next and final cluster, cluster 27. So the output just above gives us a partition derived from the hierarchy; and the table below gives us full information on the sequel, i.e., what happens next to our selected partition.

Based on these clusters, the FACOR analysis provides a cross-tabulation as follows. For clarity, we have cut the output to the first 3 factors. This detailed analysis relates to the 4 clusters above (viz., 14, 20, 22, 24). The determining influences of the factors are again seen through the correlations and contributions, only this time we are using the clusters.

```
-----------------------------------------------------------------
|CLAS ELDR YNGR| QLT WTS INR|  F#1 CO2 CTR|  F#2 CO2 CTR|  F#3 CO2 CTR|
-----------------------------------------------------------------
Representation on the factorial axes of the 3 selected nodes
   27   14   26|  01000   0|   0   0   0|   0   0   0|   0   0   0|
   26   22   25|1000 922  24| -163 922  55|  44  67   6|  11   4   1|
   25   20   24|1000 647  70|   18   3   0| 342 970 240| -35  10   5|
Representation on the factorial axes of the 4 classes of the selected partn.
   14    0    0|1000  78 284| 1933 922 647| -523  67  67| -129   4   8|
   20    3    7|1000 208 130| -217  68  22|  614 541 248| -483 335 319|
   22   10   19|1000 275 198| -590 435 213| -657 538 375|  118  17  25|
   24   21   23|1000 439  45|  130 148  16|  213 400  63|  178 278  91|
-----------------------------------------------------------------

-----------------------------------------------------------------
|CDIP ELDR YNGR| QLT WTS IND|  D#1 COD CTD|  D#2 COD CTD|  D#3 COD CTD|
-----------------------------------------------------------------
Representation on the factorial axes of the 3 selected dipoles
   27   14   26|10001000 308| 2097 922 701| -567  67  73| -139   4   9|
```

```
26    22    25|1000 922 244|  -608 263 159|  -999 709 609|   153  17  30|
25    20    24|1000 647 105|  -347 146  38|   401 194  72| -661 528 405|
```

The VACOR analysis finally gives us the cross-tabulation of these same clusters with clusters of variables. The latter are as follows.

```
Number of clusters: 8
Cluster contents follow.

Cluster j1: eBL
Cluster j2: eUS
Cluster j3: eJR
Cluster j4: eMR
Cluster j5: eSN
Cluster j6: eTG
Cluster j7: eTN
Cluster j8: eCC
```

The VACOR analysis gives the following.

	CLAS ELDR YNGR	QLT WTS INR	eBL CO2 CTR	eUS CO2 CTR	eJR CO2 CTR	eMR CO2 CTR	

Representation of the 3 nodes on the set J

```
   27    14    26|10001000    0|   32   0    0|  355   0    0|   27   0    0|  337   0    0|
   26    22    25|1000 922   24|   35   8   35|  383  75   61|   28   1    7|  350  17   25|
   25    20    24|1000 647   70|   49  75  965|  225 395  939|   16  38  993|  434 231  975|
```

Representation of the 4 classes on the set J

```
   14     0     0|1000  78  284|    0   8   99|   26  75  133|   18   1   19|  184  17   89|
   20     3     7|1000 208  130|   79  99  568|  206  90   73|    3  30  320|  332   0    0|
   22    10    19|1000 275  198|    1  37  329|  755 560  692|   57  39  628|  152 126  458|
   24    21    23|1000 439   45|   35   2    4|  234 362  101|   22   9   32|  483 551  452|
```

	CDIP ELDR YNGR	QLT WTS IND	eBL COD CTD	eUS COD CTD	eJR COD CTD	eMR COD CTD	

Representation of the 3 dipoles on the set J

```
   27    14    26|10001000 308|  -35   8  107|  -357  75  144|  -10   1   21|  -166  17   97|
   26    22    25|1000 922 244|  -48  51  553|   529 560  854|   40  42  847|  -282 167  748|
   25    20    24|1000 647 105|   44  73  340|   -28   3    2|  -19  15  133|  -150  81  155|
```

The second segment of this output table is shown in the following.

	CLAS ELDR YNGR	QLT WTS INR	eSN CO2 CTR	eTG CO2 CTR	eTN CO2 CTR	eCC CO2 CTR	

Representation of the 3 nodes on the set J

```
   27    14    26|10001000    0|   37   0    0|   61   0    0|   42   0    0|  108   0    0|
   26    22    25|1000 922   24|   40   9   80|   66  15   36|   42   0    0|   56 875  827|
   25    20    24|1000 647   70|   49  35  920|   92 135  964|   54 281000|   80  63  173|
```

```
Representation of the 4 classes on the set J
   14    0    0|1000  78 284|    0   9 147|    0  15  51|   44   0   0|  728 875 845|
   20    3    7|1000 208 130|   85  93 699|  203 478 755|   87  66 642|    4 143  63|
   22   10   19|1000 275 198|   18  12 140|    4  66 159|   14  24 350|    0 135  91|
   24   21   23|1000 439  45|   32   5  14|   40  64  35|   39   2   8|  115   4   1|
-----------------------------------------------------------------------------------

-----------------------------------------------------------------------------------
|CDIP ELDR YNGR| QLT WTS IND| eSN COD CTD| eTG COD CTD| eTN COD CTD| eCC COD CTD|
-----------------------------------------------------------------------------------
Representation of the 3 dipoles on the set J
   27   14   26|10001000 308| -40   9 160| -66  15  56|    2   0   0|  672 875 916|
   26   22   25|1000 922 244| -31  19 270| -88  91 270|  -40  27 495|  -80  42  35|
   25   20   24|1000 647 105|  54  95 571| 163 531 674|   48  65 505| -111 137  49|
-----------------------------------------------------------------------------------
```

4.4.2 Services and Other Sectors in Economic Growth

In this section, in a new study, we will look at the issue of time evolution of the variables used.

From the data of [39, 40] we retained 31 European countries. In the original data [40], 20 variables were used. We extracted the following 5 variables for our use.

LGDW	Log of initial GDP per worker. (SH)
M2	Money supply M2 as percent of of GDP, initial value. (IF)
SEC	Initial secondary school gross enrollment rate. (WB)
SRV	Per worker growth rate in Services sector. (WB)
STRD	Imports + Exports as percent of GDP, initial value. (SH)

The source of the variable is given in parentheses. The references for the sources are as follows.

IF: IMF International Financial Statistics and World Bank database.
SH: Summers, R. and A. Heston, 1991, "The Penn World Tables (Mark 5): An Expanded Set of International Comparisons 1950–88," Quarterly Journal of Economics, CVI:2, 327–68.
WB: World Bank Economic and Social Database.

Each variable has a suffix of 60, 70 or 80 denoting the decade. If the suffix is a single year it represents the value of that year (e.g., SEC60 is the secondary school enrollment rate in 1960). If the suffix is two years, then it denotes the average or the growth rate for the decade (e.g., INF6070 is the average inflation rate for the 1960s).

The variables we used were as follows.
LGDW60, LGDW70, LGDW80; M260, M270, M280; SEC50, SEC60, SEC70; STRD60, STRD70, STRD80; SRV6070, SRV7080, SRV80S.

So the data is essentially 31 countries × 5 variables × 3 decades. Our analysis is carried out on the data array crossing the 31 countries by the 15 year-related variables. Negative values in the services play havoc with the row and column masses in correspondence analysis: for this reason, we added a fixed offset. The results show very little differentiation, across time, for the GDP (gross domestic product) values, or for the services sector values. An unusual opposite trend can be noticed in regard to education, expressed as secondary school enrollments, on the one hand, and on the other hand, both M2 money supply, and import/export levels.

The R code used is as follows.

```
a <- read.table('c:/luckdata.dat')
a[,13:15] <- a[,13:15] + 0.11                    # Avoid negative values!
ac <- ca(a)

plot(ac$rproj[,1],ac$rproj[,2],type="n",
  xlab="Factor 1 (70% of inertia)",
  ylab="Factor 2 (13% of inertia)")
text(ac$rproj[,1],ac$rproj[,2],dimnames(a)[[1]],cex=0.6)
text(ac$cproj[,1],ac$cproj[,2],dimnames(a)[[2]],cex=0.6,font=2)
lines( c(max(ac$rproj[,1]),min(ac$rproj[,1])), c(0,0) )
lines( c(0,0), c(max(ac$rproj[,2]),min(ac$rproj[,2])) )
title
  ("Evolution principally of: education, imports/exports, money supply

segments(ac$cproj[1,1],ac$cproj[1,2],
        ac$cproj[2,1],ac$cproj[2,2])   #LGDW60.70
segments(ac$cproj[2,1],ac$cproj[2,2],
        ac$cproj[3,1],ac$cproj[3,2])   #LGDW70.80
segments(ac$cproj[4,1],ac$cproj[4,2],
        ac$cproj[5,1],ac$cproj[5,2])   #M260.70
segments(ac$cproj[5,1],ac$cproj[5,2],
        ac$cproj[6,1],ac$cproj[6,2])   #M270.80
segments(ac$cproj[7,1],ac$cproj[7,2],
        ac$cproj[8,1],ac$cproj[8,2])   #SEC50.60
segments(ac$cproj[8,1],ac$cproj[8,2],
        ac$cproj[9,1],ac$cproj[9,2])   #SEC60.70
segments(ac$cproj[10,1],ac$cproj[10,2],
        ac$cproj[11,1],ac$cproj[11,2]) #STRD60.70
segments(ac$cproj[11,1],ac$cproj[11,2],
        ac$cproj[12,1],ac$cproj[12,2]) #STRD70.80
segments(ac$cproj[13,1],ac$cproj[13,2],
        ac$cproj[14,1],ac$cproj[14,2]) #SRV6070.7080
segments(ac$cproj[14,1],ac$cproj[14,2],
        ac$cproj[15,1],ac$cproj[15,2]) #SRV7080.80s
```

The projections in the plane of the first two factors are shown in Figure 4.12.

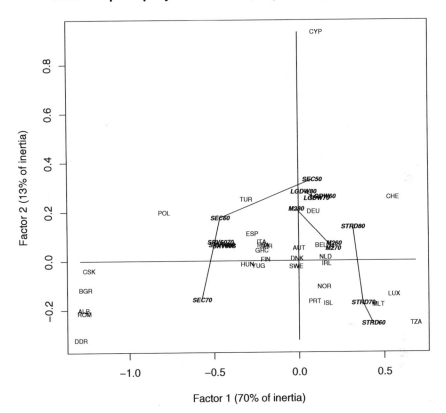

Evolution principally of: education, imports/exports, money supply

FIGURE 4.12
Factors 1 and 2 projections of 31 countries, and 5 variables each related to 3 blocks of years. The lines connect the related variables, and show the evolution.

The abstract of [39] reads: "Much of the new growth literature stresses country characteristics, such as education levels or political stability, as the dominant determinant of growth. However, growth rates are highly unstable over time, with a correlation across decades of 0.1 to 0.3, while country characteristics are highly stable, with cross-decade correlations of 0.6 to 0.9. Shocks, especially those to terms of trade, play a large role in explaining variance in growth. These findings suggest either that shocks are important relative to country characteristics in determining long-run growth, or that worldwide technological change determines long-run growth while country characteristics determine relative income levels."

Our limited analysis of this data is in agreement in that we find certain measures of growth to be inversely related to education.

4.5   Financial Modeling and Forecasting

4.5.1   Introduction

Correspondence analysis is a data analysis approach based on low-dimensional spatial projection. Unlike other such approaches, it particularly well caters for qualitative or categorical input data. Often, cluster analysis is also carried out, and the low-dimensional output representation contrasted with sets of clusters found. Input data coding impacts directly on the analysis carried out. Therefore input data coding becomes quite important when this data analysis approach is used. Examples of input data coding (see Chapter 3) include: doubling, complete disjunctive form, fuzzy coding, personal equation, double rescaling.

Our objectives in this analysis are to take data recoding as proposed in Ross [75] and study it as a type of coding commonly used in correspondence analysis. Ross [75] uses input data recoding to find faint patterns in otherwise apparently structureless data. The implications of doing this are important: we wish to know if such data recoding can be applied in general to apparently structureless financial or other data streams. Our objectives are as follows.

1. Using categorical or qualitative coding may allow structure, imperceptible with quantitative data, to be discovered.

2. Quantile-based categorical coding (i.e., the uniform prior case) has beneficial properties.

3. An appropriate coding granularity, or scale of problem representation, should be sought.

4. In the case of a time-varying data signal (which also holds for spatial data, *mutatis mutandis*) non-respect of stationarity should be checked

for: the consistency of our results will inform us about stationarity present in our data.

5. Structures (or models or associations or relationships) found in training data are validated on unseen test data. But if a data set consistently supports or respects these structures then this motivates implementation of leaving-k-out cross-validation.

6. Departure from average behavior is made easy in the analysis framework adopted. This amounts to fingerprinting the data, i.e., determining patterns in the data that are characteristic of it.

4.5.2 Brownian Motion

Efficient Market Hypothesis and Geometric Brownian Motion

The efficient market hypothesis was formulated initially by Samuelson [78]: if y_i is the value of a financial asset, then the expected value at time $t + 1$ is related to previous values as follows.

$$E\{y_{t+1} \mid y_0, y_1, \ldots y_t\} = y_t$$

When stochastic processes satisfy this conditional probability, they are termed martingales [37]. The efficient market hypothesis is taken as due to rational behavior and market efficiency. A martingale is informally a model of a fair game in that wins and loses become equal over time. An implication of the efficient market hypothesis is that price changes are not predictable from a historical time series of these prices. Empirical evidence supports the efficient market hypothesis, although Mantegna and Stanley [56] report that the additional use of fundamentals such as earnings/price ratios, dividend yields and term-structure variables allow for predictions on a longer time horizon.

Differenced values of the time series with constant time steps are studied through Brownian motion: for $0 \leq i < \infty$, the variable $y_{t+1} - y_t$ is independent of all $y_i, i < t$, and follows a Gaussian distribution. As in the efficient market hypothesis, in Brownian motion a future price depends only on the present price, and not at all on the past prices. Furthermore in Brownian motion, price difference is Gaussian. Ross [75] points to two problems with the use of Brownian motion to analyze financial data streams: firstly, use of a Gaussian implies the need for negative prices; and, secondly, it seems unrealistic to expect that a given gain or loss $y_{t+1} - y_t$ occurs with the same probability irrespective of whether y_t is large or small.

These difficulties with Brownian motion in financial time series are avoided with geometric Brownian motion. In geometric Brownian motion, the variable y_{t+1}/y_t is not dependent on any $y_i, i < t$, and $\log(y_{t+1}/y_t)$ is Gaussian. Therefore the ratio of price y_{t+1} to present price y_t follows a lognormal distribution, and is independent of all past prices. With drift μ and volatility σ, geometric Brownian motion satisfies $E\{y_t\} = y_0 \exp t(\mu + \sigma^2/2)$.

Data Transformation and Coding

Using crude oil data, Ross [75] shows how structure can be found in apparently geometric Brownian motion, through data recoding. Considering monthly oil price values, $P(i)$, and then $L(i) = \log(P(i))$, and finally $D(i) = L(i) - L(i-1)$, a histogram of $D(i)$ for all i should approximate a Gaussian. The following recoding, though, gives rise to a somewhat different picture: response categories or states 1, 2, 3, 4 are used for values of $D(i)$ less than or equal to -0.01, between the latter and 0, from 0 to 0.01, and greater than the latter. Then a cross-tabulation of states 1 through 4 for y_{t+1}, against states 1 through 4 for y_t, is determined. The cross-tabulation can be expressed as a percentage. Under geometric Brownian motion, one would expect constant percentages. This is not what is found. Instead there is appreciable structure in the contingency table.

Ross [75] pursues exploration of a geometric Brownian motion basis of Black-Scholes option cost. States-based pricing leads to greater precision compared to a one-state alternative. The number of states is left open with both a 4-state and a 6-state analysis discussed ([75], chapter 12). A χ^2 test of independence of the contingency table from a product of marginals (cf. section 2.2.1) is used with degrees of freedom associated with contingency table row and column dimensions: this provides a measure of how much structure we have, but not between alternative contingency tables. The total inertia or trace of the data table grows with contingency table dimensionality, so that is of no help to us either. For the futures data used below (see Figure 4.13), and contingency tables of size 3×3, 4×4, 5×5, 6×6 and 10×10, we find traces of value: 0.0118, 0.0268, 0.0275, 0.0493 and 0.0681, respectively. Barring the presence of low-dimensional patterns arising in such a sequence of contingency tables, we will *always* find that greater dimensionality implies greater complexity (quantified, e.g., by trace) and therefore structure.

To address the issue of number of coding states to use, in order to search for latent structure in such data, one approach that seems very reasonable is to explore the dependencies and associations based on fine-grained structure; and include in this exploration the possible aggregation of the fine-grained states.

4.5.3   Granularity of Coding

We use

1. quantile coding motivated (i) by the desire on our part to find structure in Brownian motion signals, and (ii) by the fact that it lends itself well (in that it furnishes a uniform mass density) to the analysis and display properties of correspondence analysis; and

2. an overly fine-grained set of coding categories, so that a satisfactory outcome (a *satisficing* solution in scheduling terminology) is obtained

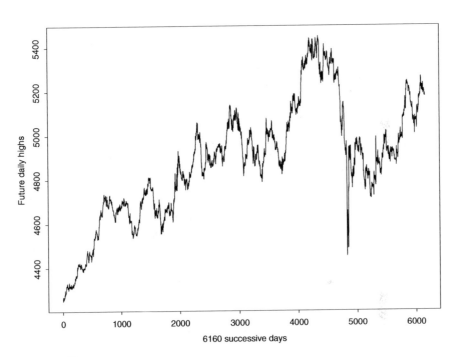

FIGURE 4.13
Future daily highs.

by aggregating these categories.

The latter objective could be sought by many different clustering approaches. Such clustering approaches are often related though: for example [87] show how data clustering through optimal graph decomposition is, in effect, one particular property of correspondence analysis. We will use correspondence analysis as an interactive analysis environment to address the following questions:

1. To aggregate the fine-resolution coding categories used, we need strongly associated coding categories.

2. Less influential coding categories are sought in order, possibly, to bypass them later in practical application.

3. In addition we will take into account possible non-stationarity over the time period of the data stream.

4. Generalizing the leaving-k-out approach to validation, we will seek consistency of results obtained for sub-intervals. If we can experimentally show that all possible sufficiently-sized sub-intervals of the time series manifest the same results, then *a fortiori* we are exemplifying how unseen data will behave.

To address point 3, and simultaneously point 4, we will take sets of 2500 values from the time series. The data to be analyzed are shown as follows. This data is derived from time series values 1 to 2500 (identifier i), values 3001 to 5500 (identifier k), 2001 to 4500 (identifier m) and values 3600 to 6100 (identifier n).

In each of these "segments" of the overall time series we consider a value and its next step value. Quantile coding is used, with 10 coding categories. So the given value in the first "segment" is coded as a 1-value in just one of the categories $i1, i2, \ldots, i10$. Meanwhile the next step value in this "segment" is coded as a 1-value in just one of the categories $j1, j2, \ldots, j10$.

The cross-tabulations that follow summarize the percentages found for each "segment" (values 1–2500; 3001–5500; 2001–4500; and 3600–6100).

Cross-tabulation of log-differenced futures data using
quantile coding with 10 current and next step price movements.
Values 1 to 2500 in the time series are used.
Cross-tabulation results are expressed as percentage (by row).

	j1	j2	j3	j4	j5	j6	j7	j8	j9	j10
i1	23.29	7.23	8.84	6.02	14.86	1.20	10.44	8.84	8.43	10.84
i2	11.60	11.60	11.20	8.80	13.20	5.20	11.60	8.80	8.80	9.20
i3	10.00	13.20	10.80	12.80	14.40	2.00	12.80	5.60	10.80	7.60
i4	8.00	9.20	9.20	12.00	15.60	4.80	12.00	10.40	9.60	9.20
i5	7.50	9.50	9.75	11.00	22.25	5.25	7.50	10.25	9.00	8.00
i6	5.05	8.08	9.09	10.10	20.20	6.06	9.09	16.16	4.04	12.12
i7	4.80	9.60	12.40	11.60	21.60	2.40	10.40	9.20	10.40	7.60
i8	8.40	7.20	8.40	12.40	13.20	7.20	8.40	10.80	11.60	12.40
i9	8.40	12.00	8.40	6.80	15.60	2.00	10.00	13.60	9.60	13.60
i10	11.20	11.60	11.60	8.00	8.00	4.00	8.80	10.00	14.80	12.00

Cross-tabulation of log-differenced futures data.
Values 3001 to 5500 expressed as percentage (by row).

	j1	j2	j3	j4	j5	j6	j7	j8	j9	j10
k1	17.67	13.65	9.24	7.22	12.05	3.21	10.44	10.84	8.03	7.63
k2	12.40	9.20	11.60	10.00	12.00	4.00	11.60	10.80	10.40	8.00
k3	12.00	11.60	11.60	9.20	18.40	3.60	12.00	6.80	6.40	8.40
k4	10.00	10.80	12.00	9.60	16.40	3.20	10.80	7.20	6.80	13.20
k5	8.78	6.59	11.46	10.98	21.71	4.63	10.24	8.54	9.27	7.80
k6	3.37	11.24	10.11	15.73	19.10	3.37	5.62	7.87	11.24	12.36
k7	7.60	11.60	7.20	9.60	16.40	4.40	9.60	12.40	12.40	8.80
k8	9.60	7.60	9.20	9.60	20.00	3.60	9.60	10.80	9.60	10.40
k9	5.60	10.00	8.80	13.20	12.80	3.20	9.60	14.00	14.00	8.80
k10	9.60	10.40	8.00	8.00	13.60	1.60	7.60	10.40	13.20	17.60

Cross-tabulation of log-differenced futures data.
Values 2001 to 4500 expressed as percentage (by row).

	j1	j2	j3	j4	j5	j6	j7	j8	j9	j10
m1	16.40	14.40	7.60	9.20	11.20	4.40	11.60	6.40	9.60	9.20
m2	11.64	10.84	12.45	10.04	12.45	4.02	9.64	10.44	10.04	8.43
m3	12.80	13.60	11.60	8.00	15.20	3.20	11.20	9.20	7.60	7.60
m4	10.40	12.80	8.80	10.40	10.00	6.00	12.40	9.60	7.20	12.40
m5	7.68	6.88	9.79	9.26	22.49	6.08	8.99	9.79	11.11	7.94
m6	7.38	8.20	9.02	12.30	22.95	2.46	6.56	9.84	9.84	11.48
m7	7.23	9.24	7.23	12.05	17.27	7.23	8.03	10.44	12.05	9.24
m8	10.80	6.80	10.40	6.80	15.20	4.80	12.40	12.00	8.40	12.40
m9	5.20	9.60	12.80	12.00	13.60	4.80	8.80	12.40	11.60	9.20
m10	10.40	8.00	10.00	11.60	11.20	4.00	8.80	10.00	12.00	14.00

Cross-tabulation of log-differenced futures data.
Values 3601 to 6100 expressed as percentage (by row).

	j1	j2	j3	j4	j5	j6	j7	j8	j9	j10
n1	18.80	14.40	7.60	8.00	8.80	3.20	10.40	11.20	8.40	9.20
n2	12.40	9.60	11.20	9.60	13.20	3.60	12.00	10.00	10.40	8.00
n3	7.97	11.55	10.76	9.96	19.92	5.18	11.55	9.16	7.97	5.98
n4	9.64	10.04	11.24	10.44	16.87	4.02	7.63	8.84	8.43	12.85
n5	9.65	7.92	12.38	9.16	21.29	4.70	11.63	6.93	9.16	7.18
n6	6.38	5.32	12.77	15.96	18.09	7.45	8.51	6.39	7.45	11.70
n7	7.60	11.60	6.80	8.00	17.60	4.40	10.80	12.00	12.40	8.80
n8	9.16	7.57	9.56	11.55	17.93	3.98	9.96	11.16	7.97	11.16
n9	4.82	10.44	8.84	12.85	15.26	1.61	9.64	12.85	14.06	9.64
n10	11.60	10.00	9.20	8.40	11.20	1.20	6.00	11.60	12.40	18.40

Figure 4.14 shows the projections of the profiles in the plane of factors 1 and 2, using all four input data tables listed above. The result is very consistent: cf. how $\{i1, k1, m1, n1\}$ are tightly grouped, as are $\{i2, k2, m2, n2\}$, reasonably so $\{i10, k10, m10, n10\}$, and so on. The full space of all factors has to be used to verify the clustering seen in this planar (albeit least squares optimal) projection.

A clustering in a full coordinate space (7 factors used) allowed a 7-cluster solution to be obtained – a solution that preceded a large increase in cluster agglomeration levels (indicated in Figure 4.15).

The clusters found are listed below. In cluster 65, coding category 9 is predominant. In cluster 68, coding categories 2 and 3 are predominant. Cluster 69 is mixed. Cluster 70 is dominated by coding category 10. In cluster 71, coding category 8 is predominant. Cluster 72 is defined by coding category 1. Finally, cluster 73 is dominated by coding category 5.

From the clustering, we provisionally retain coding categories 1; 2 and 3 together; 5; 8; 9; and 10. We flag response categories 4, 6 and 7 as being unclear and best avoided.

To check the coding relative to stationarity, Figure 4.16 shows that the global code boundaries are close to the time series sub-interval code boundaries.

The table crossing clusters (on I) and coordinates (J), giving correlations and contributions (as thousandths), is displayed in the following listing. The clusters retained here are: 65, 68, 69, 70, 71, 72, 73. Coordinates are: j1, j2, ... j10.

```
Top of hierarchy agglomerations:
( ( 65 ( 73 ( 69 71 ) ) ) ( 70 ( 68 72 ) ) )

Cluster 65: k9 n9 k7 n7 i4 m9        Predominant: 9
Cluster 68: i3 k3 m3 m4 i2 m2 k2 n2  Predominant: 2, 3
Cluster 69: n6 i8 m7                 Predominant: none
```

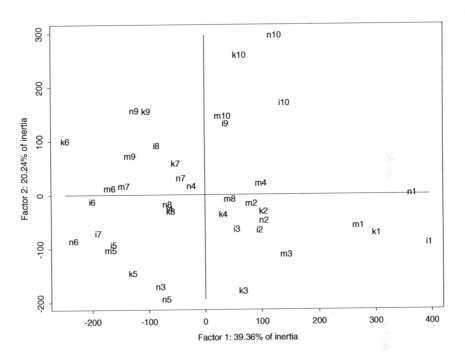

FIGURE 4.14

Factors 1 and 2 with input code categories 1 through 10 defined on 4 different spanning segments of the input data signal. Only input, or current, values are displayed here. The 4 time series sub-intervals are represented by (in sequential order) i, m, k, n. The quantile coding is carried out independently in each set of 10 categories.

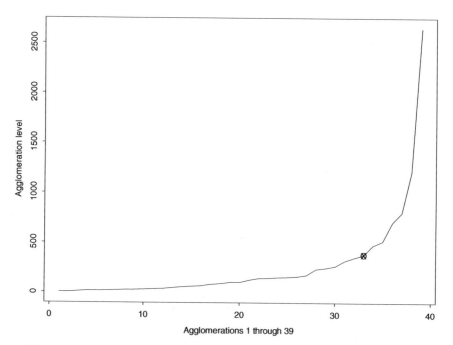

FIGURE 4.15

Cluster agglomeration levels for the hierarchical clustering (minimum variance criterion, factor projections used as input) of the 40 observations i, k, m, n in the 4 input data tables listed above.

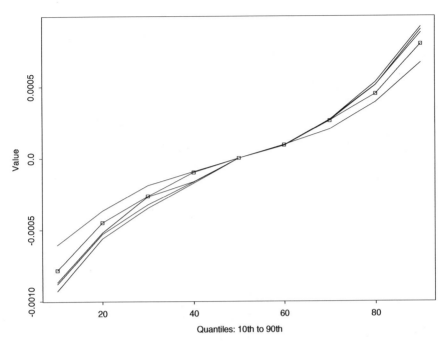

FIGURE 4.16

Stationarity of coding interval boundaries: Quantile values derived from the
4 input data tables, and – box points – averaged (by quantile).

```
Cluster 70: i10 m10 i9 k10 n10        Predominant: 10
Cluster 71: i6 k4 n4 m8 k8 n8         Predominant: 8
Cluster 72: i1 m1 k1 n1               Predominant: 1
Cluster 73: i5 m5 n3 k5 n5 k6 i7 m6   Predominant: 5
```

```
Clusters 65 through 73 represent the input coding categories.
Coordinates j1 through j10 represent the output coding categories.
```

	j1		j2		j3		j4		j5		j6		j7		j8		j9		j10	
	COR	CTR	COR	CTR	COR	CTR	COR	CTR	COR	CTR	COR	CTR	COR	CTR	COR	CTR	COR	CTR	COR	CTR
65	432	131	6	10	40	180	42	76	9	5	3	3	2	5	190	452	225	308	52	43
68	187	52	129	210	77	315	7	11	170	86	1	1	176	470	76	166	50	64	128	99
69	114	36	138	254	5	22	189	353	1	1	478	665	44	133	14	34	4	5	13	11
70	3	2	3	8	6	40	52	150	188	167	105	226	52	243	21	79	122	269	449	607
71	40	7	153	157	5	13	17	18	154	49	10	8	0	0	56	78	368	293	196	96
72	728	661	48	256	23	308	55	297	89	147	22	88	6	54	5	35	14	58	9	22
73	210	112	34	104	16	122	31	96	567	545	4	9	19	95	37	155	1	2	83	121

4.5.4   Fingerprinting the Price Movements

Typical movements can be read off in percentage terms in the input data tables. More atypical movements serve to define the strong patterns in our data.

We consider the clusters of current time-step code categories numbered 65, 68, 69, 70, 71, 72, 73 from the table of cluster results as discussed in the previous subsection, and we ask what are the likely movements, for one time step. Alternatively expressed the current code categories are defined at time step t, and the one-step-ahead code categories are defined at time step $t + 1$. Projections (e.g., Figure 4.14) are descriptive ("what is?"), but correlations and contributions point to influence ("what causes?"). Correlations and contributions are used therefore, as listed in the previous subsection, in preference to projections.

We find the following predominant movements using a threshold CTR value of 0.3 (i.e., 300 thousandths):

Cluster 65, i.e., code category 9: $\longrightarrow$ weakly 8 and more weakly 9
Cluster 68, i.e., code categories 2 and 3: $\longrightarrow$ 7
Cluster 69, i.e., mixed code categories: $\longrightarrow$ 6
Cluster 70, i.e., code category 10: $\longrightarrow$ 10
Cluster 71, i.e., code category 8: $\longrightarrow$ weakly 8
Cluster 72, i.e., code category 1: $\longrightarrow$ 1
Cluster 73, i.e., code category 5: $\longrightarrow$ 5

Consider the situation of using these results in an operational setting. From informative structure, we have found that code category 1 (values less than the 10th percentile, i.e., very low) has a tendency, departing from typical tendencies, to be prior to code category 1 (again very low). From any or all of the input data tables we can see how often we are likely to have this situation in practice: 19.04% (= average of 23.29%, 17.67%, 16.4%, 18.8%), given that we have code category 1.

In the output listing to follow we used clustered one-step-ahead (i.e., output or column) codes. We find the following predominant movements in the listing which will follow:

Cluster 65, i.e., code category 9: $\longrightarrow$ inconclusive
Cluster 68, i.e., code categories 2 and 3: $\longrightarrow$ 2, 3, 7
Cluster 69, i.e., mixed code categories: $\longrightarrow$ inconclusive
Cluster 70, i.e., code category 10: $\longrightarrow$ 8, 9, 10
Cluster 71, i.e., code category 8: $\longrightarrow$ weakly inconclusive
Cluster 72, i.e., code category 1: $\longrightarrow$ 1
Cluster 73, i.e., code category 5: $\longrightarrow$ 4, 5, 6

The table crossing clusters on I with clustered output one-step-ahead code categories j1, j14, j15, j16. Correlations and contributions are given as thousandths.

```
Top of hierarchy agglomerations:
( ( 65 ( 73 ( 69 71 ) ) ) ( 70 ( 68 72 ) ) )
```

```
Cluster 65: k9 n9 k7 n7 i4 m9        Predominant: 9
Cluster 68: i3 k3 m3 m4 i2 m2 k2 n2   Predominant: 2, 3
Cluster 69: n6 i8 m7                 Predominant: none
Cluster 70: i10 m10 i9 k10 n10       Predominant: 10
Cluster 71: i6 k4 n4 m8 k8 n8        Predominant: 8
Cluster 72: i1 m1 k1 n1              Predominant: 1
Cluster 73: i5 m5 n3 k5 n5 k6 i7 m6  Predominant: 5
```

```
Clusters 65 through 73 represent the input coding categories.
Coordinates j1 through j10 represent the output coding categories.

Top of hierarchy agglomerations for output coding categories:
( 1 ( 16 ( 14 15) ) )
```

	j1	j14	j15	j16
	j1	j2,j3,j7	j8,j9,j10	j4,j5,j6
	very low	low, spoiled	high/ very high	middle

	COR CTR	COR CTR	COR CTR	COR CTR
65|	734 131|	4 4|	260 114|	2 0|
68|	201 52|	399 583|	264 169|	137 52|
69|	210 36|	261 250|	2 1|	527 132|
70|	4 2|	26 57|	568 543|	402 229|
71|	299 7|	239 32|	17 1|	445 15|
72|	784 661|	8 37|	29 60|	179 221|
73|	277 112|	17 38|	114 113|	592 349|

From the foregoing, a possible intersection set of clusters derived from the clusters of current, and one-step-ahead future, values is:

1; 2,3; 4,5,6; ignore 7; 8,9,10.

Applying a similar fingerprinting analysis to Ross's [75] oil data, 749 values, we found that clustering the initial code categories did not make much sense: we retained therefore the trivial partition with all 10 code categories. For the output or one-step-ahead future code categories, we agglomerated 6 and 7, and denoted this cluster as 11. The output listing which follows shows the results. We find the following, generally weak, associations derived from the contributions (second of the two columns in the following listing: we used approximately 0.3, or 300 thousandths, as the cut-off value).

Input code category 6 ⟶ output code categories 1, 10 (weak)
Input code category 3 ⟶ output code category 2
Input code category 4 ⟶ output code category 4
Input code categories 9, 2 ⟶ output code category 5 (weak)
Input code category 10 ⟶ output code category 8

Not surprisingly, we find very different patterns in the two data sets of different natures used, the futures and the oil price signals.

The table crossing I with a partition of J follows. Correlation and contributions are shown. Values are given as thousandths.

	1		2		3		4		5		8		9		10		6,7	
	COR	CTR	COR	CTR	COR	CTR	COR	CTR	COR	CTR	COR	CTR	COR	CTR	COR	CTR	COR	CTR
1	208	73	7	7	225	99	85	36	239	107	9	4	9	6	211	85	7	2
2	150	143	73	208	72	86	31	36	249	303	3	4	152	269	78	85	193	138
3	47	23	287	420	335	207	42	25	14	8	163	110	59	54	49	28	4	1
4	4	4	23	86	107	166	317	473	64	101	2	3	122	281	64	91	296	274
5	221	125	0	0	152	108	115	79	71	51	139	108	126	132	137	90	38	16
6	228	295	10	39	126	204	126	197	34	57	11	19	61	148	48	71	356	345
7	347	69	126	74	105	26	29	7	2	0	154	41	147	54	11	2	79	12
8	46	26	48	82	52	37	52	36	73	53	249	195	4	5	414	274	62	27
9	5	3	6	10	55	37	43	28	460	319	48	35	5	5	6	4	373	152
10	235	237	24	73	23	29	68	83	0	0	350	481	25	46	231	268	44	33

4.5.5   Conclusions

Correspondence analysis has been shown to be a flexible, robust and scalable environment for data analysis of what presents itself initially as structureless data.

Correspondence analysis involves the alternative viewpoints offered by use of three metrics: (i) the χ^2 metric defined on profiles; (ii) the Euclidean metric, defined on factors, and far more conducive to display than the χ^2 metric; and (iii) the ultrametric, associated with a hierarchical clustering or tree representation, and permitting code category aggregation. The latter property of the analysis allows us to search for appropriate resolution level for the analysis.

We have shown that structure can be discovered in data where such structure is not otherwise apparent. Furthermore we have used correspondence analysis, availing of its spatial projection and clustering aspects, as a convenient analysis environment. Validating the conclusions drawn is always most important, and this is facilitated by semi-interactive data analysis.

5

Content Analysis of Text

5.1 Introduction

5.1.1 Accessing Content

The form of texts can certainly be analyzed statistically, but in this chapter we seek to know how much this tells us about the content of texts.

We can statistically fingerprint texts, and in studies of authorship this can be beneficial. It is well known that studies of authorship can be addressed in this way. Some work of this sort is included in the survey of correspondence analysis of text, which comprises the first half of this chapter. However a text fingerprint is rarely of equal importance in forensic science compared to a voice "fingerprint," or a shoeprint, or any one of many other author indicators. From this we draw one conclusion: text is of little use in forensics because its availability – the number of words or characters in the text – is usually very limited. The benefits of automated statistical analysis of text come from having large quantities of text.

Among other facets of statistical measurement within and between texts is encoding for (i) compression, and (ii) encryption. The former leverages on correlation within data, and the latter seeks to avoid correlation through confusion and diffusion. This allows us to draw another conclusion, namely, that in our work we want to study texts (plural) at a level of detail that is as fine and granular as possible.

We want to carry out super-resolution on our data, metaphorically going beyond the Nyquist limit, i.e., the limit of interpolation based on the given sampling. We do not want to rest with the information measure of a text data set, nor with a small number of measures of the texture of text, simply because the content of most text data sets (a chapter of a novel, part of a scientific article, etc.) cannot be reduced to this.

5.1.2 The Work of J.-P. Benzécri

The correspondence analysis tradition has been extensive and profound. We will see in our survey how the way in which textual and document analysis is carried out in the correspondence analysis approach is quite different from other contemporary and current approaches. The best general book in English on correspondence analysis is [23]. There are many books in French. For text and document analysis, see [18]. The number of articles in *Les Cahiers de l'Analyse des Données* dealing explicitly with text analysis was considerable. (If we take 3 articles out of 8 per issue, from 1980 onwards, we get over 200 articles. If we take 15 pages, on average, per article, this very hasty calculation points to a documented legacy that is impressive in its volume and potentially highly significant for theory and practice.) In the survey in the first part of this chapter, we mostly limit ourselves to going back as far as 1990. We believe that the articles we have selected incorporate adequately the experience and the findings of chronologically earlier ones.

What is even more impressive is that the themes are mostly from classical Greek and Latin philosophy and related work; Biblical and other religious texts, often in Greek and Latin; a significant number of articles on Arabic texts, including early medieval philosophy; and Russian 20th century literature. There is much in French, too, both modern and medieval, and some Spanish. Clearly there were very considerable problems to be addressed and overcome in regard to Romanized representations of character set fonts. Modern texts analyzed include free text responses in surveys, from subjects ranging from French politicians to Greek trainee teachers.

Our point of departure is an interest in the structure of discourse, pointing to the structure of argument, in text. We want to see whether such content can be found, and put under the microscope of analysis, in an automated or semi-automated way. There surely can be no better way to do this than through data analysis of the works of the father figures of thought and reason, including most of all Aristotle.

In the course of this review, we will highlight several important conclusions.

1. A metalanguage is not required as an essential to text analysis.

2. Tool words ("to," "the," "or," etc.) or grammar words identify the genre.

3. There can be no clear-cut distinction between tool words and full words. (Full words are understood here as nouns, verbs and other words conveying meaning and going beyond a simple functional role. They are used with no stemming, truncation, etc.)

4. Tool words and more content-oriented full words vary depending on the language considered. This of course is backed by experience of analysis of texts in so many different languages in the *Cahiers* articles.

5. Even if one starts with a substantial corpus one could keep expanding it by adding to it in the course of time.

5.1.3 Objectives and Some Findings

In our survey of the correspondence analysis approach to textual data analysis, as described primarily in articles in the *Cahiers*, we are mainly interested in reviewing two aspects: (i) the input data that was selected for analysis, and how it was selected; and (ii) the types of conclusions that were drawn from these studies. In between we will from time to time describe the methodology used. Since this has been fully described elsewhere in this book, we will mainly restrict ourselves for background on correspondence analysis to summary description.

Among our conclusions are the following. In regard to the sort of keywords or index terms that are most useful we find pride of place given to (i) tool words (also termed: grammar words, form words, functional words, empty words), and (ii) full words (generally with no stemming or other preprocessing). These will be discussed below. They are also relatively automatically obtained from the text. An extensive linguistic structure to support the automatic indexing is not needed.

Next, in the light of a lot of current-day work, we may ask where and how descriptive metadata and ontologies are deployed in automatic indexing or further downstream in analysis. (The word "ontology," usually used in distributed database applications in computing to mean thesaurus or concept hierarchy, is not associated in any way by us with the philosophical texts examined in the course of this review.) The surprising answer is that they are neither useful nor needed in this context. Instead a relatively basic lexical statistics approach is used to underpin the analysis.

As already noted, the articles in the *Cahiers* cover – long before 16-bit Unicode character set encoding [83] – French, Spanish, Russian (Cyrillic), Arabic, modern Greek, classical Greek and Latin. So we are not constrained to English.

Issues that we seek to cover in our survey are the practical rules of thumb needed for implementation and operational deployment. For example, we can consider a typical embedded dimensionality for use of the Kohonen self-organizing feature map when applied to textual data to be about 250. (Parenthetically, we give some details of the study of Honkela et al. [46]: using 200 Grimm tales, 250,000 words, in all, were used, of which 7000 were unique; 150 "key" words were selected, and a 270-valued coding was defined for word triplets consisting of: predecessor word, "key" word and successor word.) Or we can take the typical reduced dimensionality as targeted in latent semantic indexing (see for example [35]) to be about 70. Below we will see roughly how many index terms are used in the correspondence analysis approach, where and when cut-off thresholds are applied, what size input data set units are analyzed and how many correspondence analysis factors are used in the clustering.

5.1.4   Outline of the Chapter

In our survey, comprising the first half of this chapter, we mostly look at *Cahiers* articles in some detail. Since the title of an article is revealing of the topic, we take a non-standard approach to citation: the citation of an article usually precedes its discussion.

Section 5.2 presents a very brief outline of correspondence analysis. Section 5.2 also discusses textual data preprocessing.

In section 5.3, we review the "tool words" or "form words" (or even "empty words" or "grammatical words"), as they have been variously termed, that form a basis for a great deal of this approach to text and document analysis. We summarize results from use of a wide range of languages (with English text analyzed extremely rarely here). We discuss how a lexicon of tool words takes the place, in large measure, of a metalanguage or ontology. Since many of our later studies deal with literature sources, in section 5.3 we discuss some specific characteristics of free text survey responses.

In section 5.4 we move towards our central theme of content analysis, and how much we can achieve from the starting point of a statistical lexicon. We study: intra-document structure; the semantics of diagnosis and prognosis; the semantics of connotation and denotation; theme analysis; cognitive processes; history and evolution of ideas; stylistic expression of doctrinal themes; antinomies; and Plato's concept of The One.

In section 5.5 we return more to analysis of form. Text typology and authorship are explored, mainly in Russian and Greek literature.

A range of new case studies is then used to illustrate this methodology for free text analysis.

5.2   Correspondence Analysis

5.2.1   Analyzing Data

A very short "bird's eye" characterization of correspondence analysis will be given here.

- In this chapter we are interested in how correspondence analysis, and related hierarchical clustering, can help in interpreting a data table (array, matrix) that crosses a set of texts (or parts of texts) with a set of terms, and provides frequencies of occurrences at the intersection of text and term. Rows or columns are on occasion referred to as elements.

- Factor projections provide simultaneous display, facilitating interpretation, of the rows and columns of a data array.

- Closely coupled analyses are carried out, that are spatial – the correspondence factor analysis – and hierarchical – the hierarchical clustering.

- Correspondence analysis handles profiles, or relative category counts, exceedingly well through use of the χ^2 distance.

- In creating factor spaces, correspondence analysis maps a set of points endowed with the χ^2 metric into a set of points endowed with the Euclidean metric. In clustering these points, the mapping is continued in such a way that the Euclidean metric is mapped into an ultrametric.

- An advantage of profiles is that they can be easily cumulated or aggregated: profiles are added elementwise, with no effects on interrelationships elsewhere in the data table.

- Data normalization or standardization is accomplished through factor projections, that are then cluster analyzed.

- Active and passive analyses are supported through the practice of analyzing a principal data table, and having additional supplementary rows or columns that are projected into the factor space when the latter has been defined.

- The relationships between clusters of texts (rows) and clusters of terms (columns) may need to be taken into consideration. This is supported by interpretation aids in correspondence analysis.

5.2.2   Textual Data Preprocessing

J.-P. and F. Benzécri, "Programmes de statistique linguistique fondés sur le tri par fusion de fichiers de texte" ("Programs for linguistic statistics based on merge sort of text files"), Les Cahiers de l'Analyse des Données, vol. XIV, no. 1, pp. 59–82, 1990.

Text analysis using correspondence analysis goes back to the year 1965, in the work of B. Cordier (Escoffier) in Rennes. The following is an indicative chain of processing. Some of these operations are specific to certain languages (e.g., in regard to accents). Other operations are specific to the type of data analyzed (e.g., proper names are unlikely to be very frequent; so the likelihood of having to handle them is slim). Converting words to word forms is usually avoided, in most of the studies discussed below. In some studies, grammatical categories are used, based on words.

1. Remove accents, ligatures, punctuation.

2. Output: word, paragraph number, sentence number (within paragraph).

3. This output is sorted by word, and a count of words is obtained.

4. In other situations, the unit of analysis is a text segment, e.g., a story, report, etc.

5. The most frequent tool words are rejected, other than first or second person pronouns.

6. Words with frequency greater than a fixed threshold (here: 9) are retained.

7. Proper names are retained.

8. Objectives are analysis related to: style, literary genre, theme, author.

9. Cross-tabulation is created of words by sentence or (Biblical) verse, or chapter, or episode.

10. Words are converted to forms.

The selection of words to use always has to be settled empirically or experimentally. Firstly, there may be inflected forms – plurals to begin with. Secondly, there may be identically written words meaning different things (the verb, "book," and the noun "book"). Simplest is best, all the more so if the quantity of data analyzed is very large.

5.3   Tool Words: Between Analysis of Form and Analysis of Content

In summary, tool words (empty words) are of importance for style analysis; and full words for content analysis. However style itself can be useful for describing context, and relative properties. We will look at examples of where we can approach the analysis of content through tool words.

Full words could be rendered as, or substituted by, content words. Tool words could be rendered as form words, or grammar or grammatical words. In the survey which follows, we will summarize the assertions or conclusions. Across the many studies that we look at, if the assertions or conclusions are not always consistent, then both input data and interpretation objectives may justify this.

5.3.1   Tool Words versus Full Words

J.-P. and F. Benzécri, "Typologie de textes grecs d'après les occurrences de formes de mots. (3.D). Variantes de l'analyse globale du corpus" ("Typology of Greek texts according to occurrence of word forms. (3.D). Variants on the global analysis of the corpus"), Les Cahiers de l'Analyse des Données, vol. XIX, no. 2, 189–216, 1994.

Full words are likely to be associated mostly with particular texts. This results in a correspondence analysis producing a somewhat banal result where the first factors are easily explained by diverse texts.

Analyses taking many full words into account are of interest for structuring the themes of a corpus, especially a corpus containing many text fragments, that are related through content. Typically this could be important in document analysis. However if it is the goal to explore comprehensively an area of literature then it is best to start with tool words. Diversity of use of tool words is a very useful barometer, in many instances, of underlying content.

Very high term frequencies are truncated to a maximum acceptable threshold value. An alternative would be to require that a term must always be found in a sufficient number of texts.

H. Al Ward, "L'industrie pharmaceutique en France: analyse des réponses de huit personnalités politiques à un questionnaire" ("Pharmaceutical industry in France: analysis of the responses of eight politicians to a questionnaire"), Les Cahiers de l'Analyse des Données, vol. XVIII, no. 2, 225–238, 1993.

In this work there were 6000 forms of words, including 328 of the preposition "de" ("of," "from"), 199 of the article "des" ("of the"), etc. Since analysis of content was wanted, rather than analysis of style, such tool words or empty words are of less interest. Instead full words were used, that were sufficiently frequent. A word such as "médecins" ("doctors") did not pass the frequency threshold. In all, 26 full words were used ("assurance," "maladie," "médicale," "médicament," "médicaments," etc.).

M.-M. Thomassin and A. Hassan Hamoud, "Occurrences des mots outils et style des articles scientifiques" ("Occurrences of tool words and style of scientific articles"), Les Cahiers de l'Analyse des Données, vol. XIX, no. 1, 35–64, 1994.

Some 90 tool words were selected as being most commonly used. Rejected had been: tool words comprised of one letter (which could well be a symbol rather than a word); and a set of words that often served in a technical rather than "tool" way (examples with accents removed: "dire," "ensemble," "espace," "etat," "niveaux," etc., meaning respectively "say," "together" or "set," "space," "state," "levels"). Tool words were chosen as words that did not directly evoke the theme of the scientific article. Notwithstanding this selection criterion, theme and genre did interfere with – and even imposed themselves upon – the style.

5.3.2 Tool Words in Various Languages

J.-P. Benzécri, "Sur l'étude des textes russes d'après les occurrences des formes de mots" ("On the study of Russian texts according to occurrences of word forms"), Les Cahiers de l'Analyse des Données, vol. XIX, no. 1, 7–34, 1994.

Other than a discussion of issues in handling Russian text in a Latin encoding environment, there is a very interesting compilation of the most frequent tool words in six languages. In French, based on two different studies, one finds the following: "de," "et," "la," "les," "l'," "le," "des," "à," "en," "est," "d'," "que," "une," "dans"; and at the bottom of the list: "ne," "pas," "non." A second list, starting with nearly 34,000 occurrences was: "de," "la," "l'," "des," "les," "le," "et," "à," "d'," "en," "est," "que," "un," "une," "du," "on," "dans"; and at the bottom of the list: "ne," "pas," "non." In Latin, the decreasingly ranked list is: "et," "in," "non," "est," "ut," "quod," "ad," "qui," "cum," "sed," "si," "quae"; and at the bottom: "nec," "neque," "ne." In Spanish the corresponding list is: "que," "y," "de," "la," "a," "el," "en," "no," "con," "por," "los"; and at the bottom, "ni." Note that Spanish, with top-ranked "que," is somewhat different from French, notwithstanding what one could have expected as Romance languages. Arabic has the interesting rank order that places "an" at the top of the list, meaning the same as Spanish "que." In Greek, the rank order is: "kai," "to," "e," "de," "o"; and at the bottom of the list, "me," "ou." A rank order for Russian is also presented.

5.3.3   Tool Words versus Metalanguages or Ontologies

J.-P. and F. Benzécri, "Analyse discriminante et classification ascendante hiérarchique dans l'adjonction d'individus à un échantillon de référence: application à des données linguistiques" ("Discriminant analysis and agglomerative hierarchical classification for adding individuals to a reference sample: application to linguistic data"), Les Cahiers de l'Analyse des Données, vol. XX, no. 1, 45–66, 1995.

A major theme of current research and applied work on linguistic (textual, documentary) data relates to use of a metadescription "driver" in the form of an ontology. It is very clear though that enormous problems are immediately encountered. Firstly, automatically constructing an ontology is a difficult task, were it not for the availability of thesauri and other bibliographic support tools. Secondly, ontologies remain highly domain-specific or else lose their usefulness. Thirdly, cross-walking between ontologies is well nigh impossible beyond the domain-specific cases.

This article offers an alternative view, concluding with the following: "Thus, without having need of a descriptive meta-language, by reference alone to a sufficiently extensive corpus, the distribution of tool words revealed the genre of a text, if not its content." In other words, relative content is fingerprinted by tool words, if not absolute content. We will look now at this study.

Having already studied 889 fragments of classical Greek texts, this study seeks to place a new fragment, the treatise on the *Categories* by Aristotle. The 889 fragments comprised classical texts, and Biblical texts including the entire New Testament. The *Categories* is generally considered as the first part of the *Organon* ("tool"). The second part of the *Organon* is one of the already studied fragments. Also included among the already studied fragments is the

Isagoges by Porphyry, an introductory text that publishers since antiquity have put at the start of the *Organon*.

The *Categories*, in the state we know it, is in two discernible parts. The first deals with the concept of category. To categorize originally meant "to say something of another thing" or "to bring an accusation against someone." It may be recalled that Aristotelian terms like substance, relation, quality and quantity were not in philosophical usage but were taken from common usage. The second part of the *Categories* begins with forms of opposition, and ends with meanings of the verb "to have." Aristotle's authorship is not questioned, but the more prolix and less rigorous style raises questions about its place in relation to the *Organon*.

Based on the previous analysis of classical Greek texts, a lexicon of 200 tool words was used. Some of these were aggregated or accumulated tool words (i.e., profiles added element-wise, yielding a cumulative profile). The previous analysis provided a cross-tabulation on 889 text fragments. The present study added 19 text fragments of the *Categories* (roughly half and half from the two discernible parts of the *Categories*). The previous analysis data were used as principal elements, and the *Categories* data were used as supplementary elements.

Using 200 tool words, implying therefore a full factor space dimensionality of 200 less 1, the numbers of factors studied – the reduced dimensionality – was either 9 or 50. These analyses showed up the relationships between the 19 fragments of the *Categories* with, in particular, other works of Aristotle, of Porphyry (*Isagoges*), of Plotin (*Enneades*) and of Plato (*Phedon*). Of the 889 plus 19 fragments, a subset of 142 plus 19 were closely related. These were all philosophical texts.

Clustering of these philosophical texts showed up, firstly, a division between Plato versus all other, Aristotelian, texts. Secondly, no demarcation was verified between two discernible parts in the *Categories* – even if the two parts did appear in two clusters – implying that the two parts of the *Categories* were more unified than distinct.

Other texts were analyzed in the same way: the *On the Divine Names* by Denys (or Dionysius) the Aeropagite; the *Mystagogia* by Maximus the Confessor; the *Cratyles* dialog of Plato; book A of the *Elements* of Euclid; and *Psalms* in the Septuagint version. In all cases, factor spaces of dimensionality 9 and 50 were used. Then a hierarchical clustering was built on the principal data set. Assignment was finally carried out, using the fragments studied, with respect to clusters in the hierarchy. This discriminant analysis served its purpose well: the natural neighborhood of new fragments was obtained. The higher dimensionality of factor space, 50, performed better than the lower dimensionality, 9.

A few remarks on the tool words used in the case of Euclid's *Elements* follow. As a mathematical text, this was an outlier among the set of all texts. Coding of symbols for lines, triangles, etc. was through assimilation to proper nouns. In this document, masculine word forms are rare. Geometric terms are feminine or neuter. The most common form is the feminine article in the nominative singular.

Work along similar lines, but this time for Arabic text, is described in: *A. Chabir, "Adjonction de textes arabes à un corpus de base par l'analyse discriminante et la classification ascendante hiérarchique" ("Addition of Arabic texts to a base corpus through discriminant analysis and agglomerative classification"), Les Cahiers de l'Analyse des Données, vol. XXII, no. 3, 319–330, 1997.*

Adding description of poetry to description of classical Greek prose, used earlier in this section, is at issue in the following. Both are based on automatic indexing through tool words. This is not without its problems: a larger corpus of poetry is desirable in such analysis, since prose and poetry will necessarily be found to be different.

J.-P. and F. Benzécri, "Comparaison de textes de poètes grecs avec un corpus de prose" ("Comparison of texts of Greek poets with a prose corpus"), Les Cahiers de l'Analyse des Données, vol. XXII, no. 3, 331–340, 1997.

5.3.4   Refinement of Tool Words

J.-P. Benzécri, "Notes de lecture: langue naturelle et métalangue en analyse documentaire" ("Reading notes: natural language and metalanguage in document analysis"), Les Cahiers de l'Analyse des Données, vol. XX, no. 1, 121–126, 1995.

The most frequent full words provide the content of a text. From this point we have: (i) possible need for progressive refinement, (ii) possible need for coalescing forms of words, (iii) possible need for words appearing in all texts analyzed, (iv) documents are diverse – so that text fragments from different documents, or paragraphs within one document, may be equally worthy of exploration and investigation; (v) dominance of tool words, (vi) a sufficient starting point for analysis in the top-ranked terms or forms. Again, as already cited, "the distribution alone of tool words reveals the genre of a text, if not its content." It is clearly the case that the procedure outlined so far is based on terms found in the texts under investigation, and the automatic (or semi-automatic) indexation that results from this.

The usual approach to information retrieval is to take the user's query, and carry out the retrieval from terms in that query. The Benzécri approach is different though: retrieval should be carried out in an organized space, and not in the set of texts, albeit indexed. Admittedly, the organized space may benefit from segmentation or breaking up of the totality of the corpus. However a logical hierarchy on the part of the user in information retrieval is, in practice, a hindrance. Instead the alternative way offered by the Benzécri approach is to find the closest or otherwise most appropriate texts in the organized space.

Natural structure of language finds expression in tool words and other frequently occurring words; and the latter then allow for "the spatial analog of logical reasoning."

5.3.5   Tool Words in Survey Analysis

M. Meïmaris, "Attitudes des professeurs grecs face aux nouvelles technologies: textes des réponses libres à un questionnaire" ("Attitudes of Greek teachers relative to new technologies: free text responses to a questionnaire"), Les Cahiers de l'Analyse des Données, vol. XXI, no. 2, 221–242, 1996.

Some 650 teacher trainees answered the two following questions in free text format:

Question 1: Are you favorable to the introduction of the technologies in teaching, and for what reasons?

Question 2: If you have all technologies (television, video, computer) in class, indicate the use you are making of them. (Clarify: class, subject, topic.)

Each response was taken as two paragraphs, corresponding to the two answers. One way to code this textual data is as follows: rank all words alphabetically; classify the forms of a word (same stem, same verb, etc.) by decreasing frequency; select a lexicon of forms; and cross-tabulate, based on this lexicon, the set of paragraph pairs. Consider the lexicon as consisting of full words. Interpretation was found to be difficult. The eigenvalues decreased slowly and regularly. If an axis was easily interpreted it will be because rare words have been left in the lexicon, that are associated with a small number of subjects. On the other hand, a cross-tabulation of the lexicon with single paragraphs led to a first axis that was quite different from the following axes. On the first axis, projections of responses to question 1 – expressing a viewpoint in principle – were very different from projections of responses to question 2 – indicating specific projects. Some subjects did enter into detail on question 1, though, so the projections were not at all times clear-cut.

To further the analysis, very short responses were removed, leaving 180 subjects who had answered with at least a half page. As already described, each response is treated separately, so there were 360 "paragraphs." With, in some cases, questions not answered there were ultimately 354 paragraphs. Finally one response was dominated by a form "gnoses" ("knowledge"), and it was placed as a supplementary row so as not to overly influence the analysis. In all, there were 353 paragraphs analyzed as principal elements.

Secondly, as an external base-line reference, two texts were additionally used. One was a short autobiographical novel in 16 chapters. The other text was a set of 29 speeches in the Greek Parliament (from a debate on 25 May 1992).

First analyzed were: 180 subject responses, plus 16 novel chapters, plus 29 speeches in Parliament – in all 225 texts. Using 72 or 73 full word forms, the analysis (factor projections, classification) showed quite a clear distinction between questionnaire responses, novel chapters and Parliamentary speeches. The questionnaire responses were closer to the speeches, and relatively more distant from the lived experience expressed in the autobiographical novel data.

Therefore, we conclude that the new technologies were perceived in terms closer to oral rhetoric than to being a result of practical experience. One of the Parliamentary speeches, closer than others to the questionnaire response, was confirmed as having been made by a former mathematical teacher. One of the novel chapters based around a curriculum vitae, and correspondingly "dry" in theme, was the closest novel chapter to the questionnaire responses.

Secondly, 354 questions-responses were analyzed, using 122 word forms. Factor projections, and clustering, of responses, and word forms, were studied. The principal conclusion here was of a dimension common to the question responses, ranging from general interest through to specific projects planned for the new technologies.

5.3.6   The Text Aggregates Studied

J.-P. Benzécri, "Programme de recherches en stylométrie" ("A research programme in stylometry"), Les Cahiers de l'Analyse des Données, vol. XIII, no. 1, 97–98, 1988.

The issue of whether to select as a text object an entire book, or the work of an author, or – stepping back – a chapter or smaller unit, is an important issue. In this short summary article, the point is made that overly long text objects will be necessarily heterogeneous. If a large text, or the entirety of the works of an author, are studied, then this will appear as an undifferentiated bloc, and local particularities will be masked. Experiment alone is the final arbiter in such matters. Chapter or verse (the term used in Biblical studies for a few sentences at most) could be the most appropriate unit to use. In the case of a dialog (e.g., from the dialogs of Plato, or a play by Molière) the "role" could be the unit of analysis, that is, the speech attributed to one individual.

5.4   Towards Content Analysis

Content analysis is aided by use of full words rather than tool words. The latter may be more appropriate for analysis of style. But, as we will see, these broad generalizations must be tempered in practice.

5.4.1   Intra-Document Analysis of Content

M. Blanchet, "Analyse de la distribution des mots pleins dans l'exposé général d'une thèse sur le rapport des habitants à l'espace" ("Analysis of the distribution of full words in the text of a thesis on the relation of occupants to space"), Les Cahiers de l'Analyse des Données, vol. XX, no. 2, 225–248, 1995.

A 250-page first part of a thesis is studied. It is in the area of social psychology of the environment. Spanning the thesis, 95 consecutive fragments were considered, each containing at least 3000 characters.

As already discussed, the most frequent words in any text will be dominated by tool words – articles, prepositions, conjunctions, pronouns and so on. There are also termed empty words, i.e., empty relative to content. Tool words are useful for analysis of style, but not necessarily for analysis of content. On the other hand, non-tool full words can be very specific, and therefore not at all helpful for comparative exploration of texts or text fragments. The following procedure was therefore followed: (i) all non-tool full words with frequency greater than 49 were listed; (ii) words were added that were either tool words or non-tool words with frequency between 30 and 49; (iii) in this list, selection was made such that in 5 or more fragments, each word had to be among the 15 most frequent; (iv) a short list of words that were seemingly tool words was then purged from the list. Starting with 157 words, this procedure finally furnished 94 full words. Some exceptionally high frequency values in the resulting cross-tabulation were truncated.

An issue arises in regard to the number of factors to use as a basis for the hierarchical clustering. Using a small number of factors implies that the clustering will be based on general, methodological or thematic characters, as expressed by these factors. On the other hand, with all or nearly all the factors, far more consideration is taken of individual words. The latter will ordinarily be of greater help in seeing where the fragments analyzed are close, or distant.

Among themes brought to the fore in the analysis are: (i) relationship of occupant to environment – perception relating to physical space, changes in such perception, perception relating to social space, representation and evaluation of sociability and social practices; (ii) methodology – articulation of the object(s) studied, context, environmental psychology and previous work; (iii) socio-demographics of the Marne-la-Vallée region studied (west of Paris). All of these themes relate to the influence of man on the environment. A theme which was not found in this analysis was the influence of the environment on man.

Complementary work can be found in a later article. *M.-M. Thomassin and M. Blanchet, "Vie associative en ville nouvelle: analyse de sept entretiens avec des élus du secteur de Marne-la-Vallée" ("Social life in a new town: analysis of seven interviews with elected representatives in the Marne-la-Vallée region"), Les Cahiers de l'Analyse des Données, vol. XXII, no. 3, 341–356, 1997.*

5.4.2   Comparative Semantics: Diagnosis versus Prognosis

T. Behrakis, "Vers une analyse automatique des textes: le traitement de 42 observations continues dans les livres Épidémies I et III d'Hippocrate" ("Towards automatic text analysis: processing of 42 continuous observations in the books Epidemics I and III of Hippocrates"), Les Cahiers de l'Analyse des Données, vol. VIII, no. 4, 475–489, 1983.

The Hippocratic books, *Epidemics* I and III, contain 42 descriptions of evolution in an ill patient: changes in the face, position in the bed, movements of the hands, respiration, sweating, sleep, throwing up, state of urine, etc. Among the 42 cases, there were 25 deaths. The *Epidemics* followed an earlier work, the *Prognosis*, and sought to provide a prognosis of the general state of the patient, rather than a prognosis of the illness itself. Hippocrates terms the illnesses dealt with: phrenitis, causus and phtisie. It is still uncertain which modern terms can be applied to these illnesses.

The quantitative analysis looks at whether lexical statistics allow diagnosis of disease, or whether the state of illness, at best, is all that can be studied in this data; respectively, therefore, diagnosis of the illness, or prognosis of the general state of the patient.

The data used was the set of 42 texts crossed by the most frequent lexical forms. (A form is a sequence of letters separated by punctuation or white space.) A minimum length of 3 characters was imposed (removing most functional words – articles, prepositions, conjunctions, etc). Profiles of a few texts from *Epidemics* II, IV, V and VII were used to check the authenticity of authorship.

Analyses were carried out successively on subsamples of frequent lexical forms, starting with a minimum frequency of 10 occurrences (leading to 97 lexical forms). In this first case, classes were found corresponding to "Rapid death" with typical terms: day, middle, morning, night, calm and aggravation; a class corresponding to "Death (phrenitis-phtisie)" with typical terms: again, much, quickly, extremities, stomach, little, none, hoarse, cold, end; and a class corresponding to "Recovered" with terms twentieth (day), deafness, pain, following (days), fever. In the factor analysis, the first factor appears to be an axis of illness duration.

Subsequent analyses were based on a minimum frequency limit of 8 on occurrences of lexical forms, leading to 124 forms; and successively down to a frequency limit of 2 occurrences. Ten such analyses are summarized.

It is concluded that the principal characteristics of the data are based on recovery, death and duration of illness. From this, it is inferred that the prognosis of the general state of the patient is at issue, and not the diagnosis of the particular illness.

Analysis of other Epidemics attributed to Hippocrates showed heterogeneity of authorship, and convincing examples of different vocabulary in use.

5.4.3 Semantics of Connotation and Denotation

M. Mzabi, *"Le mot 'Joie' dans 8 romans en langue d'oïl du XIIIème siècle"* *("The word 'joy' in 8 romances in Langue d'Oïl (early French) of the 13th century")*, Les Cahiers de l'Analyse des Données, vol. VIII, no. 3, 332–342, 1983.

Based on 8 romances containing 66 fragments of text, each with one occurrence of the word "joy," context is defined in three different ways:

1. A type of questionnaire is defined relating to syntactic function (subject, direct object, etc.), distant causes of joy (finding, community, etc.), immediate causes (arrival, someone who is important, etc.), and beneficiary (king, etc.). In all, 17 questions are defined, with 99 response categories.

2. A set of 997 terms in close association with the word "joy" was defined for the set of fragments. From these, 180 terms appeared in 4 or more fragments and these were used as principal elements, with the remainder as supplementary.

3. Categories were defined for each of 13 connotations of the word "joy": time, space, society, speech, vision, religion, morality, etc. In total 36 categories were used.

These data sets are first studied individually in order to filter out the best categories, leaving others to be supplementary. Then considerable attention is given to creating a data table that integrates all data sets, with the aim of facilitating the interpretation. Interpretation is in terms of factorial axes, clusters, and relationships between fragments and terms. It is concluded that there are many and varied senses of use of the term "joy" in the fragments studied.

5.4.4 Discipline-Based Theme Analysis

S. Nodjiram and M.-M. Thomassin, *"Analyse du vocabulaire des titres des publications du Comité Français de Cartographie"* *("Vocabulary analysis of titles of publications of the French Mapping Committee")*, Les Cahiers de l'Analyse des Données, vol. XVII, no. 4, 471–480, 1992.

In this work, 800 titles of publications were used. They were organized in 29 thematic blocks, that were used in the data analysis. Words were derived from titles. Forms were used, e.g., taking singulars and plurals as one. Tool words (articles, conjunctions, prepositions, etc.) were eliminated. A lower limit of 5 occurrences overall was imposed. One exception was made for the word "zone," that occurred only 4 times. The word "Warsaw" relating to the venue of a congress, although not theme-related, was left. In all, 144 forms of words remained in the lexicon used.

S. Nodjiram, *"Analyse du vocabulaire mathématique des titres des articles de deux périodiques: un siècle du Bulletin de la S.M.F. et du Journal de Liouville"* *("Mathematical vocabulary analysis of the titles of articles from two journals: a century of the Bulletin of the SMF [Société Mathématique de France] and of the Journal de Liouville"), Les Cahiers de l'Analyse des Données, vol. XVII, no. 2, 133–158, 1992.*

From the two publications, 37,600 words were used. About 4000 article titles are used, following selection in regard to short communications, letters to the editor, etc. The most frequent words were tool words: "sur," "de," "des," "la," "les," "d'," etc.; then full words: "fonctions," "équations," "théorie," "surfaces," "théorème"; further down the list were verb forms: "concernant," "être," "peuvent." The word list used consisted of 241 mathematically relevant full words ("point," "points," "poisson," "polaires," etc.).

5.4.5   Mapping Cognitive Processes

M.C. Noël-Jorand, M. Reinert, S. Giudicelli and D. Drassa, "Discourse analysis in psychosis: characteristics of hebephrenic subject's speech," Proc. JADT 2000: 5^{ièmes} Journées d'Analyse Statistique des Données Textuelles.

While this work is not associated with the *Cahiers de l'Analyse des Données*, it is based on Benzécri's work. Used are content words, i.e., words that are amenable to analysis, and function words, i.e., words used as supplementary elements to help describe the classes found with the content words. The majority of the words used were verbs, nouns, pronouns and speech markers. It is noted that the language production of the hebephrenic patient seems to be very limited. The hebephrenic patient who is a schizophrenic subject exhibiting negative symptoms produced a specific type of verbal behavior and speech pattern, that were clearly distinguished from the positive symptom patients.

5.4.6   History and Evolution of Ideas

T. Behrakis and E. Nicolaïdis, "Typologie des prologues des livres grecs de sciences édités de 1730 à 1820: humanisme et esprit des Lumières" ("Typology of prologs of Greek books on science published between 1730 and 1820: Humanism and the spirit of the Enlightenment"), Les Cahiers de l'Analyse des Données, vol. XIV, no. 1, pp. 9–20, 1990.

Thirty-four books containing about 600 pages of prolog material were considered. The Greek national revolution took place in 1821, so that the prolog material deals with the sciences and with learning, on the one hand, and with some elements related to nationality on the other. Although correspondence analysis is used in this work, the greater part of the discussion relates to the major clusters of prologs and authors.

In the prologs the most frequently occurring lexical forms were considered: 793 forms appeared 10 times or more. A length constraint was also applied: lexical forms were at least 3 characters long. Hierarchical clustering led to the top part of the hierarchy being used for the analysis. This consisted of six clusters. Cardinalities and indicative labels were: 18 prologs and 1 prolog in clusters characterized as "spirit of Humanism"; a cluster of 9 prologs characterized as "spirit of the Enlightenment"; and clusters of 2, 1 and 3 prologs characterized as attempts to write mathematics in an approachable way.

The "spirit of Humanism" clusters dealt with mathematics, were influenced by ancient Greek traditions, tended towards use of archaic language, stressed the debt owed by Europe to Greek classicism, and were published early in the period covered in publishing houses across Europe associated with the Greek diaspora. These "spirit of Humanism" clusters contained 18 prologs, and had 21,734 associated lexical forms of which 8,522 (39.2%) were distinct.

The "spirit of the Enlightenment" clusters dealt with chemistry, physics and astronomy, in approachable language. They were more taken up with the potential of the nation, and the role of language and education for the future of the nation. These prologs were from books published towards the end of the period covered, and were exclusively published in Vienna. The "spirit of the Enlightenment" clusters contained 9 prologs, and had 10,202 associated lexical forms of which 2,230 (21.9%) were distinct.

Various particular terms are studied (methodos = method, theorema = theorem, episteme = science, glossa = language, Europa = Europe, genos = race, ethnos = nation, idees = ideas, erga = works, peiramata = experiment, neoi = young people, technes = arts or techniques, monas = unit, metron = measure, arithmos = member, stoicheia = elements, and so on). Use of genos in preference to ethnos is discussed. Various characterizing elements are in evidence: the tendency among "spirit of Enlightenment" authors to use science for national revival and to relate this to European progress in science and technology. The tendency among "spirit of Humanism" authors is to seek national revival by linking up with a glorious past, and this is close to the spirit of religious humanism in the Greek Orthodox church in the 18th and 19th centuries. Faced with these important orientations, perhaps one regrettable tendency in the "spirit of Enlightenment" was for mathematics to become somewhat separated from the sciences.

5.4.7 Doctrinal Content and Stylistic Expression

C. Rutten and J.-P. Benzécri, "Métaphysique d'Aristote et métaphysique de Théophraste: analyse comparative des chapitres fondée sur les fréquences d'emploi des parties du discours" ("Metaphysics of Aristotle and of Theophrastus: comparative chapter analysis based on frequencies of use of parts of speech"), Les Cahiers de l'Analyse des Données, vol. XIV, no. 1, 37–58, 1990.

Not much of the work of Theophrastus who followed Aristotle exists now, although he gave occasion to the word "Lyceum" which he headed, whereas Plato before him had given us the word "Academy." Theophrastus was a center of concern to the 17th century French literary figure, La Bruyère, and figured in the latter's philosophical and botanical works.

In this study, 142 chapters of the *Metaphysics* of Artistotle are analyzed together with 9 chapters of his follower, Theophrastus. The data are comprised of the distribution (i.e., frequency counts) by chapter of ten grammatical categories (adjectives; adjective-pronouns, pronouns and numerals; adverbs; articles; coordination conjunctions; subordination conjunctions; particles; prepositions; substantives; and verbs).

A first analysis was based on the Aristotle data as principal data, and the Theophrastus data as supplementary. However it was preferred to treat all as principal. The first factor had a chronological interpretation, and the second was dominated by the adjective category. In-depth analysis of clusters of (fragments of) chapters follows. Theophrastus was seen to be stylometrically close to Aristotle when dealing with themes already covered by the latter.

An in-depth study of particular clusters of text passages is then carried out. We have the following from the conclusion to this article. The ideal of stylometric research would be achieved if we could always establish a reversible link between the statistical properties of the texts and the *raison d'être* of these properties, in terms of genre and meaning.

In the study of the *Metaphysics* of Artistotle and the *Metaphysics* of Theophrastus, the chapters belonging to one class were found to be very much characterized by frequency of adjectives. Most of these chapters set out a dualist doctrine. These chapters bear witness to a concern to characterize the antithetical aspects of the real.

It is precisely adjectives, if necessary in the form of substantives, more than any other part of speech that serve to express such properties. We can consider the opposition of the intelligible and the sensible, or the incorruptible and the corruptible, or the immobile and the mobile, or good and evil (bad).

Thus, in this work, a link can be drawn between stylistic expression (based on grammatical categories) and doctrinal content.

C. Rutten and J.-P. Benzécri, "Analyse comparative des chapitres de la Métaphysique d'Artistote fondée sur les fréquences d'emploi des parties du discours: confrontation entre l'ordre du textus receptus, les références internes et l'ordre du premier facteur" ("Comparative analysis of the chapters of the Metaphysics of Aristotle based on the frequencies of use of parts of speech: confrontation between the order of the textus receptus, internal references, and the order on the first factor"), Les Cahiers de l'Analyse des Données, vol. XIII, no. 1, pp. 41–68, 1988.

The *textus receptus* is the way that Aristotle's *Metaphysics* appears today, having been subject to an unknown number of rewritings, omissions and changes over the centuries. The question posed is whether the internal references in the 142 chapters point to a different order to that used in the *textus receptus*. If chapter 9 in book Z, for example, makes a supposition in regard to chapter 7, then a directed link is defined from Z7 towards Z9. An in-depth discussion follows on orders of chapters, referred to the first factor. The study concludes with an order that results from this discussion.

5.4.8   Interpreting Antinomies Through Cluster Branchings

J.-P. and F. Benzécri, "Typologie de textes grecs d'après les occurrences de formes de mots. (3.C) Analyse de l'ensemble du corpus" ("Typology of Greek texts using occurrences of word forms. (3.C) Analysis of the entire corpus"), Les Cahiers de l'Analyse des Données, vol. XIX, no. 2, pp. 171–188, 1994.

Using Greek classical, Hellenist and Biblical sources, 889 text fragments were crossed with 132 forms of tool words. In some cases cumulated tool words (i.e., clusters) were used, and in such cases the original tool words were linked into the analysis as supplementary elements. The number of factors used as a basis for the clustering does not have any strict rule: experimentation is needed. With a small number of factors, leading to general structure being brought out, one could expect that the upper part of the hierarchical clustering is clearly interpretable. With a large number of factors, more local affinities between text fragments would be given higher standing.

Characterizations of clusters include the following. The characterization is mostly carried out by examining branches. The two way split in a dendrogram lends itself well to this. Thus, in correspondence analysis interpretation of results [23], cluster dipole analysis plays an important role.

1. Strict formalization, typical of texts on logic.

2. Content related to man, versus content related to being.

3. Arguments based on rhetorical rather than dialectical lines: hence, recitations and exhortations, or "images in prose."

4. History versus fable.

5. Exhortation versus laudation.

6. Homilies (Virgin Mary, female word forms).

7. Generalization and analogy, expressing and exemplifying mysteries.

8. Collectivities (plural word forms).

5.4.9   The Hypotheses of Plato on The One

L. Brisson and J.-P. Benzécri, "Structure de la seconde partie du Parménide de Platon et répartition des vocables" ("Structure of the second part of Plato's Parmenides and distribution of the terms"), Les Cahiers de l'Analyse des Données, vol. XIV, no. 1, pp. 117–126, 1990.

The *Parmenides*, written by Plato around 369 BC, is a hypothetical dialog involving Parmenides, Socrates and Zeno. A first part deals with the doctrine of intelligible forms, used by Socrates to criticize the arguments of Zeno, but then the doctrine is critiqued by Parmenides. Following more debate that turns to intelligible forms rather than sensible things, Parmenides agrees to give a demonstration of his standpoint of The One. In the second part of the dialog, the positive and negative consequences are examined of whether The One is, or is not, for one, and for others. The combinations considered by Parmenides are in the form of a branching. At the root is The One. On the next level are the two nodes: "if it is," and "if it is not." Each of these have two subnodes, "positive consequences" and "negative consequences." Finally each node at that level has subnodes "for self" and "for others." Parmenides is dialoging with a young man by the name of Aristotle (not the later great philosopher) who was to become one of the tyrants of Athens. The discourse of Parmenides, only, is taken into consideration.

Now, from Plotin (who died in 270 BC) and likely from well before, it has been considered that certain lines of the argument of Parmenides actually contain an additional hypothesis. The hypothesis relating to the following is considered to contain two hypotheses in fact: The one; if it is; positive consequences; for self. (In other words, this is one branch of the tree, from root = "the one," through to the terminal node = "for self.") This hypothesis that we consider therefore to split in two will be labeled 2, with a possible split into 2a and 2b. The number of words relating to the different hypotheses, based on the corresponding lines of the discourse of Parmenides, are as follows: 1 – 1636; 2a – 5031; 2b – 531; 3 – 692; 4 – 402; 5 – 1152; 6 – 329; 7 – 515; and 8 – 250.

Whether hypothesis 2 ought or not to be taken as hypotheses 2a and 2b has been debated in the literature using philosophical and philological arguments. In the analysis of the relevant texts, it is shown that there is no stylometric basis for subdividing 2 into 2a and 2b.

In each of the 8 fragments of discourse, termed "hypotheses," there were 32 words. A principal 32×8 table was analyzed, and the fragments 2a and 2b were taken as supplementary. Fragments 2 and 2a were found to be more or less coincident, which is consistent with their associated total number of words (cf. above). Fragment, or sub-hypothesis, 2b is very close also. This points to little support in this data for subdividing hypothesis 2 into sub-hypotheses.

The analysis of this data is described further in section 5.9.

5.5   Textual and Documentary Typology

5.5.1   Assessing Authorship

G. and A. Volchine, "Étude comparée de textes russes: le Don Tranquille et d'autres oeuvres de M.A. Cholokhov; les nouvelles de F.D. Kriukov" ("Comparative study of Russian texts: 'Quiet Flows the Don' and other works of M.A. Sholokhov; and short stories by F.D. Kriukov"), Les Cahiers de l'Analyse des Données, vol. XX, no. 1, 7–26, 1995.

Authorship between works, between chapters, and even within chapter, are studied in the scope of the great work of Sholokhov, one of many on the Don Cossacks between the First World War, through the Civil War, and up to the period of collectivization. Other works were included also: *Virgin Soil Upturned, They Fought for Their Country* and *The Fate of a Man.* For contrast with this work, five short stories by F.D. Kriukov were included also: in total a corpus of 747,000 words. *Quiet Flows the Don* consists of four books each subdivided in eight parts, with each again divided in chapters. It was written at an uneven tempo between the years 1928 and 1940, and comprises 393,500 words. *Virgin Soil Upturned* was written in 1932 (40 chapters) and 1959 (29 chapters), deals with collectivization and the opening up of virgin territories in the Don region, and comprises 214,500 words. *They Fought for Their Country* was written in 1956-1957 and comprises 604,000 words in 26 chapters. *The Fate of a Man* is different, it relates to a personalized account of World War II, and it comprises 106,000 words. The short stories of Kriukov were published between 1903 and 1913, other than one (*Kazatchka*) published in 1896, comprise 68,000 words, and relate to the customs and practices of the Don Cossacks.

To avoid thematics, the following words were not taken into account: proper nouns, substantives, qualifying adjectives or adjectives of appearance, personal pronouns, verbs with the exception of "to be" and "to become," and numerals with the exception of "one" due to its interesting forms and meanings.

Retained were: prepositions, adverbs, adverbial forms of adjectives, conjunctions, demonstrative pronouns and possessive pronouns. These parts of speech were, by and large, the most frequent in any case.

The units of text used ranged from the 8 parts of *Quiet Flows the Don* through to a single unit for each of *The Fate of a Man* and the short stories of Kriukov: in all, 17.

The number of word forms ultimately retained was 224. These were found in all texts (other than *The Fate of a Man* which, stylistically, as already noted, was very different). Mapped back onto words, these 224 forms were related to 236,283 words in *Quiet Flows the Don*, i.e., 32% of the total 747,000 words. It is noted that these tool words comprise 32.4% of the earlier chapters, and 43.6% of the last chapters. Other works of Sholokhov are up to 44%. For Kriukov, the percentage is more constant, around 36%.

Analysis is carried out on 224 tool word forms, crossed by 17 texts. After preliminary analyses, a subset of 142 tool words was used. This subset was such that each tool word had an overall frequency of an arbitrary threshold of 235 occurrences across all texts. In some later analyses the five Kriukov short stories were considered as one text, giving in total 13 texts.

In all of the principal factor plots from correspondence analysis, and in the clustering, the interesting fact emerged that parts 1 through 6 of *Quiet Flows the Don* were counterposed to parts 7 and 8, and the latter were close to other Sholokhov texts. The Kriukov texts (considered separately or as one unit for analysis) are clearly separated. To put more under the microscope the lack of continuity in *Quiet Flows the Don*, a detailed analysis of chapters was undertaken. Given the paucity of forms or words, the following approach was used. All texts, as in the analysis up to now, were used with the exception of the part or short story, etc. under investigation; and for that part or short story, the tool word form frequency profiles are entered into the results as supplementary elements.

The analysis is characterized by use of forms of tool words. Were personal pronouns to be used, it is remarked, then they would have introduced proximities that would have falsified totally the graphical outcome. Of particular pertinence here is *The Fate of a Man* that was written in the first person. The use of tool words counteracted any over-influence arising from this.

This work is characterized by the size of the corpus used. It was not limited to sequences of characters or word forms. Homonyms were resolved, and where necessary compound words were resolved.

A perplexing finding from this work is that parts 1 through 6 of *Quiet Flows the Don* are very different from remaining parts, and from other works by Sholokhov. It is as if these were due to a different author. Solzhenitsyn thought as much. There had been allegations of plagiarism against Sholokhov, at least for the first three parts of his major work, and indeed it has been suggested that these were taken from Kriukov. However this analysis shows clearly that, whatever about the issue of plagiarism, these early parts of *Quiet Flows the Don* are very unlike the works of Kriukov.

J.-P. and F. Benzécri, "Sur les comparaisons entre deux corpus analysés, d'abord, séparément" ("On the comparison between two corpora analyzed, at first, separately"), Les Cahiers de l'Analyse des Données, vol. XX, no. 1, 67-78, 1995.

This study takes up the large-scale Sholokhov and Kriukov study, together with a wide range of other short texts – from Lenin, 17th century texts and poems. It is concluded that a set of some 160 forms of tool words appears to suffice for constructing a general taxonomy of texts in literary language, across genres. However authorship is more an open issue, on the basis of such a restricted, albeit carefully selected, set of forms.

G. and A. Volchine, "Typologie de textes russes d'après l'emploi des parties du discours dans les trois mots initiaux de chaque phrase: applications à des oeuvres relatives aux Cosaques du Don" ("Typology of Russian texts based on use of parts of speech in the first three words of each sentence: applications to works related to the Don Cossacks"), Les Cahiers de l'Analyse des Données, vol. XXII, no. 4, 421-442, 1997/8.

This work precedes by a number of years the work already surveyed in this section.

The authors tackle the very large volume of textual data from Sholokhov and Kriukov, but this time they examine the grammatical categories of words at the start and at the end of each sentence. Statistically, in writing just as with the human voice, an individual begins and ends sentences in relatively typical ways, giving rise to a statistical signature for that individual. The data consisted of the text fragments crossed by a triplet of parts of speech given by the first three words of a sentence.

Sentences were delimited by any punctuation signs, with the exception of commas and dashes (both used in a particular way in Russian). The grammatical encoding adopted was based on solid grounding but did admit of alternatives. The following were used: adjectives; adverbs – of adjectival form, certain gerunds usually taken as adverbs, and adverbs issuing from pronouns; conjunctions – simple (coordination, subordination), pronouns used as conjunctions, particles used in this way, and adverbs used in this way; gerunds – present, past; cardinals; pronouns – personnel, used as particles, and others; prepositions – adverbs used in this way, and all other cases; particles – used as adverbs, and not; participles; negations; substantives – in the nominative, and otherwise; dashes; verbs – imperfect, and perfect; commas and other sentence non-separators; and finally hybrids – conjunctions with adverbs and particles, or with pronouns and adverbs.

There were 92 grammatical triplets, defined in this way, that were present at least twice in the corpus. To give an idea of the coding process, the first part (of eight in all) of *Quiet Flows the Don* had 2520 triplets; the second part had 2817 triplets; the third part had 3877 triplets; and so on. The number of textual units used was 12.

As in the previous study using tool words, it was again very noticeable how parts 1 through 6 of *Quiet Flows the Don* were very different from parts 7 and 8. With the latter were other works by Sholokhov. Again, Kriukov was closer to the latter, but somewhat distant.

A study using 135 different pairs of grammatical categories starting sentences was also carried out, with a fairly similar outcome. However the triplets are considered to be more robust, in particular for analysis of linguistic categories.

Results were also reported on where both triplets and the tool words of the previous study were simultaneously analyzed. Corroboration was obtained through independent analysis, and through conjoint analysis, of the two studies using different input data.

It is concluded that at least the first three chapters of Sholokhov's major work were not due to him. However it is also noted that the ordering of parts 1 through 8 is by and large chronological. It is not entirely excluded that the author changed his "signature" in a very significant way, over the decades through which the quiet Don flowed.

J.-P. Benzécri, "Philologie classique et taxinomie des textes" ("Classical philology and textual taxonomy"), Les Cahiers de l'Analyse des Données, vol. XVIII, no. 3, 377–382, 1993.

Is sentence length a measure of style? Sentence length is bound up with syntactic complexity, and is not a good unit of study in its own right. On the other hand, tool words can capture many aspects of syntactic complexity. Consider the following: different types of conjunctions and relative pronouns;

counts by mode of verbal forms; nouns by case; or in languages without noun inflection then nouns that are subjects of verbs, nouns that are direct complements, and nouns governed by a preposition. Stemming and truncation are not always necessary.

A prosaic assessment is that: "... literary creation attests to the fact that the greatest works are like lightening propagating along a contingent line between the high potential of a longtime charged spirit on high, and the earth, which the spirit thereby burns..."

J.-P. Benzécri and M. Meïmaris, "Comparaison globale entre les oeuvres de deux auteurs Platon et Xénophon" ("Global comparison between the works of two authors, Plato and Xenophon"), Les Cahiers de l'Analyse des Données, vol. XXI, no. 4, 403–430, 1996.

The digital repository, Thesaurus Linguae Grecae (TLG), from the University of California at Irvine, is used, not without problems of font, diacritical marks, etc. A first analysis cross-tabulates 35 works from Plato against 80 forms of tool words. In this analysis it was found useful to omit personal pronouns, which were somewhat dominant, but to retain the singular neuter article, and conjunctions "and" and "otherwise." Subsequent analysis compared the philosopher and the historian named in the title of this article. There is a short discussion on M. Clay, "Typologies syntaxiques des phrases de trois textes anglais scientifiques de niveaux différents" ("Syntactic typologies of sentences of three scientific texts of different levels, in English"), in [18]. Here, an author wrote three different texts, one for the peer-reviewed literature, one as a technical description for practitioners, and thirdly one for the public and published in a newspaper. Variables were used such as word length, use of the passive and subordination of propositions. The three levels of text could be very clearly distinguished on this basis.

In this article the practical issue is raised as to whether stylometry requires a new set of tool words whenever the corpus is extended. The answer is no: the same lexicon can well be sufficiently stable to allow use on extended or enhanced data.

5.5.2   Further Studies with Tool Words and Miscellaneous Approaches

J.-P. and F. Benzécri, "Typologie de textes grecs d'après les occurrences de formes de mots" ("Typology of Greek texts using occurrences of word forms"), Les Cahiers de l'Analyse des Données, vol. XVIII, no. 2, 143–176, 1993.

From classical and Biblical sources, 250,000 words are studied, as full words and as tool words.

J.-P. Benzécri and M. Meïmaris, "Comparaison entre les oeuvres de trois auteurs grecs classiques: Platon, Xénophon et Thucydide" ("Comparison between the works of three classical Greek authors, Plato, Xenophon and Thucydides"), Les Cahiers de l'Analyse des Données, vol. XXII, no. 1, 7–12, 1997.

(Title and theme noted here only.)

J.-P. and F. Benzécri, *"Typologie de textes espagnols de la littérature du siècle d'or d'après les occurrences des formes des mots outil"* (*"Typology of Spanish texts from the literature of the Golden Age, using occurrences of tool word forms"*), Les Cahiers de l'Analyse des Données, vol. XVII, no. 4, 425–464, 1992.

(Title and theme of this Spanish-language study are noted here only.)

J.-P. and F. Benzécri, *"Typologie de textes latins d'après les occurrences des formes des mots outil"* (*"Typology of Latin texts using occurrences of tool words"*), Les Cahiers de l'Analyse des Données, vol. XVI, no. 4, 439–466, 1991.

J.-P. Benzécri, *"Typologie de textes grecs d'après les occurrences des formes des mots outil"* (*"Typology of Greek texts using occurrences of tool words"*), Les Cahiers de l'Analyse des Données, vol. XVI, no. 1, 61–86, 1991.

(Title and theme of these Latin- and Greek-language studies are noted here only.)

A.M. Alkayar, *"Classification d'un ensemble varié de textes français d'après les occurrences de mots pleins"* (*"Classification of a varied set of French texts based on occurrences of full words"*), Les Cahiers de l'Analyse des Données, vol. XVIII, no. 2, 239–244, 1993.

A total of around 22,000 words is used, from 25 text fragments, spanning the time period from the 18th century to the present day. It is indicated that the diversity of texts aids greatly in discrimination between genres.

J.-P. Benzécri and A. Chabir, *"Essais pour une stylométrie appliquée aux textes arabes"*, Les Cahiers de l'Analyse des Données, vol. XIII, no. 1, 69–80, 1988.

Twelve Arabic texts or groups of texts with 10,000 words were used. Characterization was in terms of 95 categories of parts of speech ("partes orationis arabicae").

Y.-L. Cheung, S.-C. Leung and J.-P. Benzécri, *"Essai typologique des écritures manuscrites chinoises"* (*"Typological experiment on Chinese hand-written manuscripts"*), Les Cahiers de l'Analyse des Données, vol. XIII, no. 1, 81–92, 1988.

Chinese ideograms are characterized in terms of width and height, symmetry, tilt and slant.

J.-P. Benzécri, *"Sur l'étude des textes arabes d'après les occurrences des formes de mots"* (*"On the study of Arabic texts based on occurrences of forms of words"*), Les Cahiers de l'Analyse des Données, vol. XIX, no. 1, 65–84, 1994.

The problems to be addressed in automatic indexing of Arabic text are discussed. An initial study is carried out of the commentary by the early medieval philosopher Averroes (Ibn Roshd) on Book Z of Aristotle's Metaphysics. Some other Arabic texts of around the same period are used for comparison.

J.-P. Benzécri and A. Chabir, *"Typologie d'un ensemble de textes arabes d'après les occurrences de formes de mots et de locutions"* (*"Typology of a *set of Arabic texts using occurrences of forms of words and phrases"), Les Cahiers de l'Analyse des Données, vol. XX, no. 4, 389–412, 1995.*

Classical and modern poetry are analyzed, based on 91 text fragments. Automatic indexing of word forms and phrases (defined as character sequences delimited by white space or punctuation) are used.

É. Évrard, "Analyse factorielle de la composition métrique de l'haxamètre dactylique latin" ("Factor analysis of metrical composition of the dactyl hexameter in Latin"), Les Cahiers de l'Analyse des Données, vol. XIII, no. 1, 9–18, 1988.

The analysis is on 25 works by a range of authors crossed by 16 hexameter metrical forms.

J. Denooz, "Application des méthodes d'analyse factorielle à la fréquence des catégories grammaticales en latin" ("Application of factor analysis methods to frequencies of grammatical categories in Latin"), Les Cahiers de l'Analyse des Données, vol. XIII, no. 1, 19–40, 1988.

The analysis is on 42 works and 10 different grammatical categories.

5.6 Conclusion: Methodology in Free Text Analysis

The first studies surveyed below are related to analysis of texts crossed by carefully selected terms. The objective is to throw light on important associations between text and term. Exogenous variables (such as sex, age or national origin) can be easily incorporated passively into the analysis.

L. Lebart, "Une procédure d'analyse lexicale écrite en langage Fortran" ("A procedure for lexical analysis written in Fortran"), Les Cahiers de l'Analyse des Données, vol. VI, no. 2, 229–243, 1981.

As an example five poems on the theme of autumn (the fall) are analyzed, from: Lamartine, Gauthier, Verlaine, Brugnot and Baudelaire. Brugnot and Gauthier are closely placed in the factor analysis output, with others well separated. The simultaneous displays of authors and terms aid in interpretation.

A. Salem, "La typologie des segments répétés dans un corpus, fondée sur l'analyse d'un tableau croisant mots et textes" ("Typology of repeated segments in a corpus, based on the analysis of a table crossing words and texts"), Les Cahiers de l'Analyse des Données, vol. IX, no. 4, 489–500, 1984.

Slogans and other repeated expressions are studied, derived from 16 revolutionary press sources from Paris in the summer of 1793. Some expressions ("all the..." or "all these...") typified certain news sheets. Syntagmatic (i.e., phrase) analysis permits analysis of vocabulary changes in a political or other group. Given the fixed period at issue here, the objective was to study authorship and typology of texts.

H. Akuto and L. Lebart, "Le repas idéal: analyse de réponses libres en trois langues: anglais, français, japonais" ("The ideal meal: analysis of free responses in three languages: English, French and Japanese"), Les Cahiers de l'Analyse des Données, vol. XVII, no. 3, 327–352, 1992.

The words used to characterize each text subject were not exclusively food names, but they were in their great majority. The Japanese data was from 1008 subject responses. There were 832 distinct word forms (following Romanization), and of these 139 were present a sufficient number of times (i.e., 7) in the set of responses. The French data was from 1000 subject responses. There were 1229 distinct word forms, of which 112 word forms were present sufficiently frequently ($\geq$ 18 occurrences). It was noted how singular and plural of word forms used were often not in the same clusters: this points to the benefits of refraining from word stemming. Age groups, and sex, were used as supplementary variables. Much of the discussion was based around differences between geographic regions, and these two subject properties of age and sex.

J.-P. Benzécri, "Notes de lecture: sur l'analyse des réponses libres dans une enquête internationale" ("Reading notes: on the analysis of free text responses to an international survey"), Les Cahiers de l'Analyse des Données, vol. XVII, no. 3, 353–358, 1992.

An interesting discussion is presented, following the Tokyo-Paris-New York data analysis. It is pointed out how Romanization of the Japanese text has its pitfalls. Furthermore, there is full agreement that blind word stemming is potentially counter-productive: it is a "costly operation" in terms of deleting potentially useful information. An argument is then presented against statistical tests, and instead an approach is favored which uses other, additional counter-balancing and contrast-rich data. This is an important point, since it is very different from current practice. For example, the work of Noël-Jorand et al. [69] discussed in section 5.4.5 above, using the Alceste software package (I-Image Society), is based on a significant frequency of occurrence using the χ^2 statistic. We have seen examples of the Benzécri alternative here: adding more textual data; using a data set as principal relative to other data sets as supplementary, or vice versa; and use of discriminant analysis to place a textual data set, or its component fragments, among an already-built textual spatial structure.

Full words refer to nouns, adjectives, verbs and adverbs, and represent *grosso modo* objects, qualities, actions in time and categories of these actions. Tool words are other words – conjunctions, prepositions, pronouns, etc. – which provide glue or linkage between the full words, or play the role of being generic substitutes for the full words.

The distinction between full word and tool word is not always crystal clear. In practice a lexicon is defined as follows. A list of all word forms found in the data under study is made, and is ranked by frequency of occurrence. Word forms occurring once only (hapax legomenon: thing said once only, in the Bible, and thus causing difficulty for interpretation) or infrequently are removed. From what remains on our list, the following is done in order to retain only tool words. All nouns (including all inflected forms in the case of a language like Latin), verb form, and pronoun form related to the first or second person are removed.

From the tool words, the author's style can be recognized, and also the genre of the text: poetic, historic, rhetorical, didactic, etc. Style and genre are closely related.

From the full words, access is provided to the theme of the text. The situation relating to first and second person forms is peculiar: an affirmation of presence by the writer points in the direction both of genre (e.g., in the theater) and also content.

There had been some discussion in Akuto and Lebart (see above in this section) of tool words. In principle, however, style analysis is not an objective in survey research, so tool words ought not be used in such a case.

In work on Greek, Latin, and so on, usually a text fragment contained at least 500 words. In the survey, however, each response contained less than 20 word occurrences on average. Because of the small size of the response texts, a very sparse cross-tabulation matrix resulted. There are ways to avoid this, however. Consider the lexicon used as L, having (say) 100 word forms. Consider then a set M of categories of word forms, comprising (say) 20 categories. The table crossing L and M will be analyzed. Each of the (say) 1000 surveyed subjects, set I, have measurements in L. So put I as a set of supplementary elements: $L \times (M \cup_s I)$. This table is then analyzed. Alternatively, we can use discriminant analysis, assigning each subject i to the closest category m. If we consider target categories M' based on e.g., combined age-sex-profession, then the cross-tabulation of results in the form of $M' \times M$ will show how close the data are to our assumptions.

Absence of words in the subject responses can also be included easily. This gives rise to a type of "virtual" word.

It is also interesting to consider the analysis, for some J, of the data sets from the three cities, and hence languages, leading to an analysis of $(M_1 \cup M_2 \cup M_3) \times J$. This would allow us to see if "fromage" was really perceptually near to "cheese."

Further possibilities for analysis of conjoined and derived data tables is provided. In all of this, we see that the analysis is at a fine level of granularity, but simultaneously the analysis must avoid being deluged in data. The fine level of granularity is crucial for detailed analysis of content.

5.7   Software for Text Processing

Software discussed here, and some of the data sets, are available on the web at: www.correspondances.info .

A few programs in the C language are available for preparing text for correspondence analysis. We take each text, or each part of a text, as separate text files. On the book's web site there is a collection of individual aviation accident reports (analyzed below) in a set of text files. The names of these files are listed in the file `filelist.txt`.

Running the file `txtanalysis` with two input parameters, viz., the file with the list to be processed, `filelist.txt`, and the output file name, defaulted to `words.txt`, produces a ranked list of all words (demarcated by white space or punctuation marks) in the texts analyzed. For the aviation report files used, we find a word list of length 4238 words.

This list of words, with default file name `words.txt`, can be refined or modified using any editor. We created a new list of 35 words, in the file `words0.txt`, from the words "pilot" (9th ranked word originally) down to "control" (42nd ranked word originally).

A frequency cross-tabulation of texts versus words is created by the program `xtabulate`. It is run as: `xtabulate words0.txt filelist.txt` and an output file name that is defaulted to `xtabulate.txt`. The latter file is set up to be read by `read.table` in R.

In the case where different collections of texts are under investigation, we may wish to have a consistent word list that is shared by all. Program `word_analysis` takes as input: a file listing the superset of words; a file listing the text file names; a threshold which is the necessary number of occurrences of words in each text file. The output produced, then, is the words that are common, in sufficient numbers, to each text. This word list can be used as input to the cross-tabulation program.

In the case of Aristotle's *Categories*, the word list is produced using program `docanalyse`. It is run in this way: `docanalyse arist10x.txt words.txt`. The file `arist10x.txt` is the entire text of the *Categories* [4]. Manual processing can then be carried out on the file `words.txt`, in order to retain selected terms.

The cross-tabulation in this single document situation is made with program `doctabulate`. Use: `doctabulate arist10x.txt words.txt out`. This produces not one but a set of cross-tabulations, or contingency tables, `out1.txt`, `out2.txt`, `out3.txt`, `out4.txt`, corresponding to the different section levels in the *Categories*.

5.8 Introduction to the Text Analysis Case Studies

Up to now, in this chapter, we have reviewed the use of (i) tool words, or empty or grammar words, and (ii) content words, or full words, to study inter- and intra-relationships between or in sets of texts. These sets of texts could be: book chapters; event reports; narrative stories; etc. An important consideration is that such blocks of text be similar in length, which we specify as roughly the same order of magnitude in numbers of words. It was clear too that a general definition of "word" is needed, which we took as contiguous characters, without distinction of upper case and lower case, nor grammatical category (e.g., singular or plural, or verb form) and with white space or punctuation used as demarcation.

Such a framework for studying one or more texts leads to a statistical lexical perspective. This allows us to characterize any given word by all of its contexts. We can assert that a word is located at the center of gravity of its contexts. Such a perspective is a central one in correspondence analysis (cf. the formulas linking the dual spaces, which permits simultaneous display of rows and columns in the data matrix analyzed). This is a statistical perspective in the sense of an ensemble perspective.

It can be easily distinguished from a very different type of perspective (perhaps, but not necessarily, based on an interrelationship graph), that seeks or concentrates on salient, or authority or hub, words. This latter vantage point is of greater interest for web querying, for example. Thus, with Google or similar search engines, we seek texts (or other objects) that cite specific words, and furthermore we want retrieved texts ranked in order of importance (or relevance, or authority). Having salient or authority words helps in the structuring of text, in that hyperlinks can be established between XML-tagged words. Much work has been carried out, and continues to be, with this perspective.

Following this discussion, we suggest that the statistical lexical approach to inter-text analysis is most appropriate for forensics, and comparative evaluation (e.g., relating to authorship). For intra-text analysis, the statistical lexical approach offers a quantitative means of characterizing meaning, or semantics. We will now look at the potential (and limitations) of semantic analysis of text through lexical statistics on the basis of a number of case studies.

5.9   Eight Hypotheses of Parmenides Regarding the One

Earlier in this chapter, we looked as the dialog of Parmenides regarding the Platonic One. We list in Table 5.1 the data used, and show some of the analysis results.

From Figure 5.1, hypothesis 2a more or less coincides with 2. This is not unexpected since 2b is quite brief, thereby allowing the other component of hypothesis 2, viz. 2a, to be clearly seen as tantamount to 2. Furthermore, 2b is not at all far from 2 nor 2a. This leads to the suggestion that hypothesis 2 is not justifiably subdivided into 2a and 2b, at least not from the textual

TABLE 5.1

Thirty-two terms crossed 8 fragments. The fragments are related to hypotheses in the discourse of Parmenides, concerning the Platonic One. The additional two supposed sub-hypotheses of hypothesis 2, 2a and 2b, are also given.

	h1	h2	h2a	h2b	h3	h4	h5	h6	h7	h8
allos	20	134	131	3	21	23	24	8	24	13
an	47	123	114	9	17	7	27	7	8	3
anomoios	4	14	10	4	4	4	6	1	3	1
ara	35	72	71	1	4	11	20	6	3	4
autos	60	149	142	7	15	13	27	12	14	4
gar	10	37	32	5	9	5	12	13	6	4
ge	29	54	49	5	13	3	14	4	7	3
de	30	126	117	9	15	7	26	8	11	2
dé	5	33	26	7	7	2	8	3	8	1
eautou	42	77	77	0	11	4	4	1	4	1
ei	17	43	41	2	8	8	25	4	7	7
eimi	137	398	371	27	55	25	127	30	42	28
<eis	63	255	240	15	48	28	45	4	21	16
>en	28	95	87	8	3	8	5	1	4	3
eteros	34	80	80	0	5	4	9	2	9	1
ego	18	37	34	3	5	6	12	2	3	1
kai	54	328	285	43	38	17	57	3	38	9
lego	3	21	20	1	1	3	24	3	2	4
men	15	48	44	4	6	4	8	1	5	1
mé	10	51	45	6	4	1	69	13	10	12
mén	22	28	26	2	5	1	10	3	2	2
o	132	659	625	34	78	44	120	28	41	21
omoios	5	15	11	4	3	4	1	1	3	1
oti	6	18	16	2	3	1	6	1	1	2
ou	31	92	81	11	11	4	31	4	8	3
oukoun	7	31	30	1	5	1	5	1	5	2
oute	73	84	63	21	1	13	6	18	3	6
outos	12	52	45	7	9	4	10	7	4	2
te	9	93	76	17	14	4	13	1	8	4
tis	29	44	40	4	6	3	16	5	5	5
tís	5	24	24	0	3	1	6	4	3	1
faino	3	15	15	0	1	2	4	1	12	5

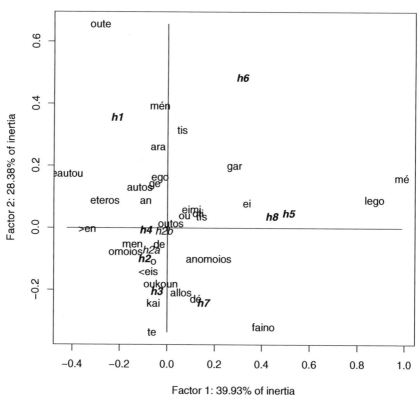

FIGURE 5.1

Principal factor plane of Parmenides data. The possible component hypotheses of h2, viz. h2a and h2b, are taken as supplementary elements. All other hypotheses, 1–8, are taken as principal elements.

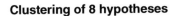

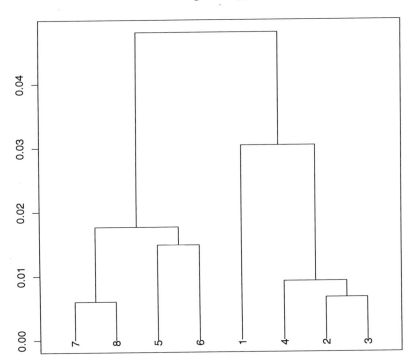

FIGURE 5.2

Clustering of 8 hypotheses derived from the Parmenides data.

evidence available here.

Figure 5.2 leads to taking into account either two clusters of the hypotheses, or all of the hypotheses individually.

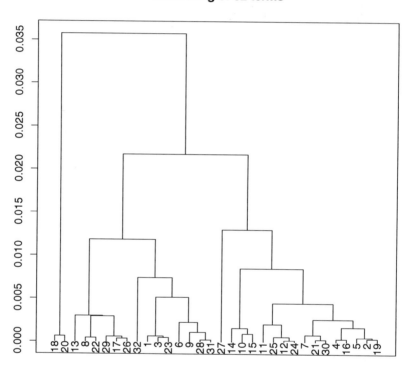

FIGURE 5.3

Clustering of 32 terms used in analysis of Parmenides data.

TABLE 5.2
Cluster identifiers used in the following tables.
Two clusters of hypotheses (cf. Figure 5.2),
and 16 clusters of terms (cf. Figure 5.3).

Cluster identifiers	Terms
Cluster i6:	gar
Cluster i11:	ei
Cluster i13:	<eis
Cluster i14:	>en
Cluster i27:	oute
Cluster i32:	fainó
Cluster i36:	ara, egó
Cluster i39:	de, o
Cluster i41:	autos, an, men
Cluster i42:	ou, eimi, oti
Cluster i43:	te, kai, oukoun
Cluster i44:	legó, mé
Cluster i45:	allos, anomoios, omoios
Cluster i46:	ge, mén, tis
Cluster i47:	eautou, eteros
Cluster i48:	dé, outos, tís

Cluster identifiers	Hypotheses
Cluster j13:	h7, h8, h5, h6
Cluster j14:	h1, h4, h2, h3

There follow projections, correlations and contributions of 16 clusters of terms crossed by two clusters of hypotheses, labeled j13 and j14. See Table 5.2 for memberships of all clusters.

	CLAS	ELDR	YNGR		QLT	WTS	INR		j13	CO2	CTR		j14	CO2	CTR	
6	0	0	1000	15	12		365	773	34		635	227	34	
11	0	0	1000	18	14		361	773	41		639	227	41	
13	0	0	1000	74	7		179	773	21		821	227	21	
14	0	0	1000	23	19		88	773	53		912	227	53	
27	0	0	1000	31	6		162	773	16		838	227	16	
32	0	0	1000	7	23		512	773	66		488	227	66	
36	4	16	1000	37	0		213	773	1		787	227	1	
39	8	22	1000	208	12		191	773	33		809	227	33	
41	5	38	1000	96	6		188	773	17		812	227	17	
42	25	33	1000	164	11		266	773	31		734	227	31	
43	29	37	1000	115	5		195	773	14		805	227	14	
44	18	20	1000	36	205		593	773	583		407	227	583	
45	1	40	1000	52	2		255	773	5		745	227	5	
46	7	35	1000	48	1		243	773	1		757	227	1	
47	10	15	1000	44	27		108	773	77		892	227	77	
48	9	34	1000	33	2		266	773	6		734	227	6	

There follow projections, correlations and contributions of 16 clusters of terms crossed by the 8 hypotheses, h1 through h8. See Table 5.2 for memberships of clusters.

	CLAS	ELDR	YNGR		QLT	WTS	INR		h1	CO2	CTR		h2	CO2	CTR		h3	CO2	CTR	
6	0	0	1000	15	48		104	37	9		385	74	40		94	27	26	
11	0	0	1000	18	24		143	4	0		361	260	70		67	0	0	
13	0	0	1000	74	30		131	59	9		531	12	4		100	329	199	
14	0	0	1000	23	29		190	53	8		646	202	67		20	183	109	
27	0	0	1000	31	133		358	485	315		412	36	54		5	100	271	
32	0	0	1000	7	74		70	31	11		349	35	30		23	19	28	
36	4	16	1000	37	24		222	352	41		456	73	20		38	140	68	
39	8	22	1000	208	38		120	293	55		582	383	166		69	6	4	
41	5	38	1000	96	17		196	511	43		515	0	0		61	14	5	
42	25	33	1000	164	22		164	38	4		477	137	35		65	1	0	
43	29	37	1000	115	65		94	310	98		605	221	162		76	22	29	
44	18	20	1000	36	260		56	63	80		312	82	240		22	31	162	
45	1	40	1000	52	58		86	198	56		484	11	7		83	30	35	
46	7	35	1000	48	39		256	643	121		403	224	98		77	16	13	
47	10	15	1000	44	54		264	491	130		545	12	8		56	10	11	
48	9	34	1000	33	19		103	219	20		509	0	0		89	104	40	

h4	CO2	CTR	h5	CO2	CTR	h6	CO2	CTR	h7	CO2	CTR	h8	CO2	CTR
52	7	6	125	1	0	135	827	323	62	7	4	42	20	15
67	99	40	210	396	37	34	1	0	59	9	2	59	229	84
58	140	71	94	105	12	8	307	75	44	14	5	33	33	15
54	26	13	34	358	41	7	109	26	27	60	20	20	8	4
64	23	52	29	121	63	88	190	207	15	44	66	29	1	1
47	1	1	93	4	1	23	1	1	279	706	586	116	204	234
71	259	107	134	19	2	33	3	1	25	142	38	21	14	5
38	9	6	108	44	7	27	22	7	39	105	45	17	137	81
39	5	1	100	138	9	32	3	0	43	35	7	13	294	78
28	217	82	154	551	48	33	8	1	48	4	1	31	43	15
29	42	47	100	42	11	7	252	134	68	90	65	20	21	21
17	14	62	403	690	699	69	50	105	52	0	0	69	71	285
92	430	429	92	43	10	30	0	0	89	205	134	45	82	74
22	79	52	128	5	1	38	17	5	45	5	2	32	11	6
28	26	24	45	286	61	10	83	37	45	3	2	7	89	74
33	21	7	112	6	0	65	512	79	70	107	23	19	30	9

From Table 5.2 and the above factor analysis output listing, we see, for example, how cluster 44 (legó, mé) is strongly associated with hypothesis 5; how cluster 27 (oute) is positively associated with hypothesis 1; how cluster 32 (fainó) is strongly associated with hypotheses 7 and 8. We can continue in this way, characterizing all clusters in their (mutual) importance for the set of hypotheses. Proceeding further, we can examine these mutual influences using clusters of hypotheses.

5.10 Comparative Study of Reality, Fable and Dream

Our objectives in the following include a study of whether tool words allow us to differentiate between (i) technical texts and literature in the form of a novel; (ii) literature in the form of a novel, and fairy tales; (iii) dream reports and accident reports; (iv) dream reports and literature in the form of a novel; (v) dream reports and fairy tales; (vi) different novels by the same author; and (vii) chapter texts from a novel, and sub-chapter texts. We will also investigate content or full words from each group of texts. As we will see, typically we take a text to be very approximately 1000 words in length, as an arbitrary but satisfactory standard. We selected about 1000 texts from sources that covered the above areas.

5.10.1 Aviation Accidents

The NTSB aviation accident database [10] contains information from 1962 about civil aviation accidents in the United States and elsewhere. We selected 50 reports. Examples of two such reports used by us: occurred Sunday, January 02, 2000 in Corning, AR, aircraft Piper PA-46-310P, injuries – 5 uninjured; occurred Sunday, January 02, 2000 in Telluride, TN, aircraft: Bellanca BL-17-30A, injuries – 1 fatal. In the 50 reports, there were 55,165 words. Report lengths ranged between approximately 2300 and 28,000 words.

Sample of start of report 30: *On January 16, 2000, about 1630 eastern standard time (all times are eastern standard time, based on the 24 hour clock), a Beech P-35, N9740Y, registered to a private owner, and operated as a Title 14 CFR Part 91 personal flight, crashed into Clinch Mountain, about 6 miles north of Rogersville, Tennessee. Instrument meteorological conditions prevailed in the area, and no flight plan was filed. The aircraft incurred substantial damage, and the private-rated pilot, the sole occupant, received fatal injuries. The flight originated from Louisville, Kentucky, the same day about 1532.*

5.10.2 Dream Reports

From the Dreambank repository [36, 38, 79] we selected the following collections:

1. "Alta: a detailed dreamer," in period 1985–1997, 422 dream reports.

2. "Chuck: a physical scientist," in period 1991–1993, 75 dream reports.

3. "College women," late 1940s in period 1946–1950, 681 dream reports.

4. "Miami Home/Lab," in period 1963–1965, 445 dream reports.

5. "The Natural Scientist," 1939, 234 dream reports.

6. "UCSC women," 1996, 81 dream reports.

To have adequate length reports, we requested report sizes of between 500 and 1500 words. With this criterion, from 1 we obtained 118 reports, from 2 and 6 we obtained no reports, from 3 we obtained 15 reports, from 4 we obtained 73 reports, and finally from 5 we obtained 8 reports. In all, we used 214 dream reports, comprising 13696 words.

Sample of start of report 100: *I'm delivering a car to a man – something he's just bought, a Lincoln Town Car, very nice. I park it and go down the street to find him – he turns out to be an old guy, he's buying the car for nostalgia – it turns out to be an old one, too, but very nicely restored, in excellent condition. I think he's black, tall, friendly, maybe wearing overalls. I show him the car and he drives off. I'm with another girl who drove another car and we start back for it but I look into a shop first – it's got outdoor gear in it - we're on a sort of mall, outdoors but the shops face on a courtyard of bricks. I've got something from the shop just outside the doors, a quilt or something, like I'm trying it on, when it's time to go on for sure so I leave it on the bench. We go further, there's a group now, and we're looking at this office facade for the Honda headquarters.*

With the above we took another set of dream reports, from one individual, Barbara Sanders. A more reliable (according to [38]) set of reports comprised 139 reports, and a second comprised 32 reports. In all 171 reports were used from this person. Typical lengths were about 2500 up to 5322. The total number of words in the Barbara Sanders set of dream reports was 107,791.

5.10.3   Grimm Fairy Tales

We used all 209 tales from the Brothers Grimm [70]. In all, in the 209 text files, there were 280,629 words. Story lengths were between 650 and 44,400 words.

Sample of start of tale 9, "Rapunzel": *There were once a man and a woman who had long in vain wished for a child. At length the woman hoped that God was about to grant her desire. These people had a little window at the back of their house from which a splendid garden could be seen, which was full of the most beautiful flowers and herbs. It was, however, surrounded by a high wall, and no one dared to go into it because it belonged to an enchantress, who had great power and was dreaded by all the world. One day the woman was standing by this window and looking down into the garden, when she saw a bed which was planted with the most beautiful rampion – rapunzel, and it looked so fresh and green that she longed for it, and had the greatest desire to eat some.*

5.10.4   Three Jane Austen Novels

The 61 chapters of *Pride and Prejudice* [8] were used as separate texts. In all, there were 122,809 words. Typically chapters contained between 4350 and 29,357 words.

The 24 chapters of *Persuasion* [9] were used as separate texts. In all there were 83,559 words. Typically chapters contained between 1579 and 7007 words.

The 50 chapters of *Sense and Sensibility* [7] were used as separate texts. In all there were 120,491 words. Typically chapters contained between 1028 and 5632 words.

We wondered if text length, alone, could influence comparative properties of a group of texts. To see if this had an influence, we additionally took the 50 chapters of *Sense and Sensibility* and broke them into sub-chapters of lengths between about 500 and 1000 words. We deleted sub-chapters with less than 500 words. We were left with 131 sub-chapter texts, containing in total 113,058 words. How we use these sub-chapters will be described in the next section.

5.10.5   Set of Texts

In all we used 910 texts, comprising 931,537 words. A word list extracted from all texts yielded 20,452 unique words. (Examples, with frequencies of occurrence: the 53,179; and 40,543; to 31,842; of 23,244; in 15,503; it 15,382; was 15,043; ... zany 1; zen 1; zeros 1; zest 1; zigzag 1; zippers 1; zipping 1; zmed 1; zoe 1; zulu 1. We recall that a "word" is a set of contiguous characters, with punctuation replaced by white space, and with previous conversion of all upper case characters to lower case.) We considered the following groups of texts:

1. 50 aviation accident reports

2. 171 dream reports (Barbara Sanders, batch 1: 139, batch 2: 32)

3. 214 dream reports (Alta, college women, Miami home/lab, and natural scientist)

4. 209 tales of the Brothers Grimm

5. 61 chapters from Jane Austen's *Pride and Prejudice*

6. 24 chapters from Jane Austen's *Persuasion*

7. 50 chapters from Jane Austen's *Sense and Sensibility*

8. 131 sub-chapters derived from Jane Austen's *Sense and Sensibility*

5.10.6   Tool Words

The correspondence analysis approach to analyzing text makes use of tool words (form words, grammatical words, [semantically] empty words) as opposed to content words (full words). The former are often parts of speech and appear to convey little if any semantic content. Tool words can be useful for stylometry. More than that, however, the style used is often a good measure of content. Importantly, since tool words are shared between texts from very different domains (with the sole assumption that relatively standard English is used), they allow for inter-text analysis. On the other hand, content words are very likely to be domain-specific, so that common aspects of texts are not likely to be apparent. Clearly content words are a far better basis for intra-domain analysis. Investigating major factors in aviation accidents, for example, could only be approached in this way. Using content words, it is likely that evolution of dream themes, or chronology changes in the works of the Brothers Grimm or of Jane Austin, would appear to be best addressed, although stylistic changes could additionally be sought on the basis of tool words.

Our objective here is inter-text and inter-domain analysis, and so tool words are most appropriate. We note again that words are defined as follows: all cases were set to lower case; and all punctuation marks were replaced by a blank (word separator). A contingency table was created that cross-tabulated frequencies of occurrence of these tool words in the texts. Correspondence analysis involves mapping a χ^2 distance defined on *profiles* of contingency table row vectors, and column vectors, into a Euclidean factor space.

Labels used in the figures to follow (Figures 5.4, 5.5 and 5.6) are:

1. 50 aviation accident reports: denoted a.

2. 171 dream reports (Barbara Sanders, batch 1: 139, batch 2: 32): denoted b.

3. 214 dream reports (Alta, college women, Miami home/lab, and natural scientist): denoted d.

4. 209 tales of the Brothers Grimm: denoted g.

5. 61 chapters from Jane Austen's *Pride and Prejudice*: denoted 1.

6. 24 chapters from Jane Austen's *Persuasion*: denoted 2.

7. 50 chapters from Jane Austen's *Sense and Sensibility*: denoted 3.

8. 131 sub-chapters derived from Jane Austen's *Sense and Sensibility*: denoted 4.

From a total of 20,452 words found in all of these 910 texts, we used the 500 top-ranked words. Figure 5.4 shows a clear separation of the technical texts relating to aviation accidents (a). All four sets of the Austen novels are overlapping.

Figure 5.5 shows the same projection plot, but with the aviation accident reports not displayed (these were used in the definition of the factors), with the aim of showing more clearly the interrelationships between the remaining data sets.

Figure 5.6 restricts what is displayed still further, to focus on the dream reports only. In particular we found here that the Barbara Sanders reports, labeled b, were more compactly positioned compared to the other reports, labeled d.

5.10.7 Domain Content Words

To go beyond the tool words used thus far, we carried out the following procedure.

Factors 1 and 2: 910 reports or chapters, 500 common terms

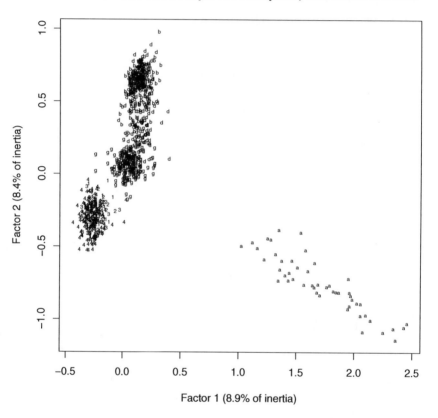

FIGURE 5.4

All 910 texts, crossed by the 500 most common words, shown in the principal factor plane.

Factors 1 and 2: 860 reports/chapters (aviation reports omitted)

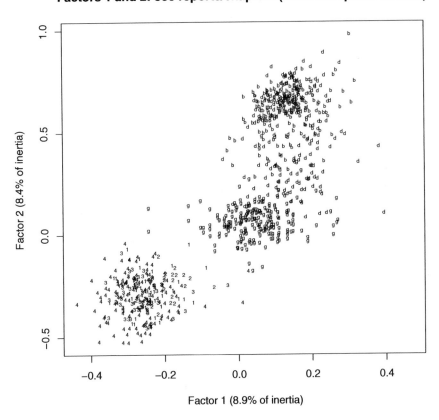

FIGURE 5.5

Display of 860 texts (all other than the technical, aviation accident ones), crossed by the 500 most common words, shown in the principal factor plane.

Factors 1 and 2: 385 dream reports

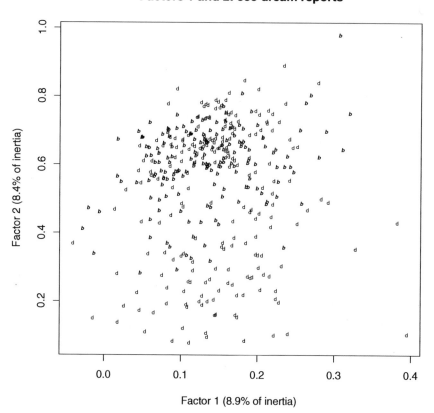

FIGURE 5.6

Display of 385 dream reports, crossed by the 500 most common words, shown in the principal factor plane.

- Determine the global set of words, in all texts, as done so far, and call this w. Let $w[1 : 500]$ be the 500 top-ranked words. We characterize $w[1 : 500]$ as definite tool words.

- We seek content words instead therefore in the set $w[501 : 1500]$.

- These content words are sought in a domain-limited set of documents.

- Further, we require that a content word is to appear at least a minimum number of times. This threshold was set at 10.

With this procedure we found, for the aviation accident reports, 161 content words. Some of these are as follows: appeared 21; length 12; hard 10; aircraft 212; several 18; case 19; remained 35; change 18; forced 17; runway 200.

For the dream reports, we found 584 content words. Examples: living 81; best 28; fine 35; inside 138; forth 25; deal 49; hands 73; met 24; appeared 23; dead 40; death 16.

For the Brothers Grimm tales, we found 585 content words. Examples: carriage 62; living 23; master 175; best 54; fine 70; inside 71; forth 158; deal 24; hands 83; met 80; appeared 49; castle 209.

Finally for the three Jane Austen novels, we found 745 content words. Examples: carriage 110; living 82; master 34; best 112; fine 89; wentworth 218; forth 21; deal 112; hands 44; met 89; appeared 87; dead 15; death 32.

Note how these words are certainly not unique to their corresponding domains. For this reason, we prefer to characterize them as content-oriented words.

5.10.8   Analysis of Domains through Content-Oriented Words

Figure 5.7 displays aviation accident reports, based on content-oriented words. Looking at extreme reports 33, 39 and 41, these relate to, respectively: collision on aborted take-off; near collision when taxiing on runway; and damage sustained in hard landing. Reports elsewhere in this display related to loss of power in flight, forced landings, flight into terrain, striking cable, etc. We will not look further at this small sample of accident reports that are primarily used here as a baseline collection of technically written texts.

Content-oriented display of dream reports can be seen in Figure 5.8. Very good discrimination was found in the first factor plane. Symbol – is used for the more reliable set of 139 reports from Barbara Sanders. A further set of 32 less reliable reports by the same respondent are labeled o. All of these Barbara Sanders reports are quite well separated from all other dream reports. Symbol | is used to label reports from "Alta, a detailed dreamer" from the period 1985–1997. These are fairly well separated from the remaining dream reports used, 96 in all, from: "college women," "Miami home/lab," and a "natural scientist." Among the latter 96, we found some possibility of discriminating in the first factor plane, but our numbers of reports are small (respectively,

50 aviation accident reports, 161 domain terms

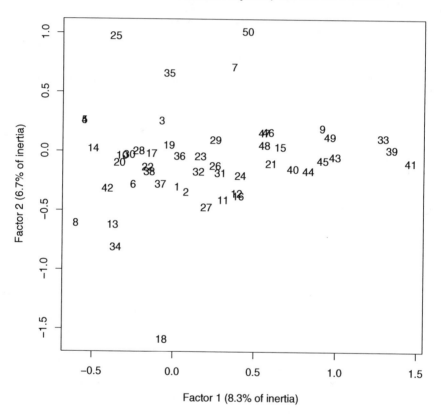

FIGURE 5.7

Principal factor plane of 50 aviation accident reports, crossed by 161 content-oriented words.

15, 73 and 8) so we will not push this issue any further. It may be noted that the first factor plane accounts for about 2.7% of the total inertia of the data. (Depending on whether we look at the rows – reports – or columns – words used, we have a full space dimensionality of either 385 or 584, less one given the data coding implied by use of the χ^2 distance in correspondence analysis).

The analysis of the Grimm Brothers tales was unusual: almost all tales were near the origin, signaling very average profiles. However tale number 162 was distant from the origin on the (negative) vertical axis; and tale number 24 was distant from the origin on the (negative) horizontal axis. Axes are not invariant to sign so that fact is irrelevant. What is relevant is that these two tales were found to be very different from all others, and this was the case for different reasons. We found tale 24 to be highly characterized by occurrences of the word "Gretel"; and tale 162 to be highly characterized by occurrences of the word "Frederick."

The Jane Austen novels (see Figure 5.9) are nicely distinguished in the principal factor plane. We recall that we use, respectively, the 61 chapters of *Pride and Prejudice* labeled by symbol 1, the 24 chapters of *Persuasion* labeled by symbol 2 and the 50 chapters of *Sense and Sensibility* labeled by symbol 3.

5.11 Single Document Analysis

5.11.1 The Data: Aristotle's *Categories*

From earlier in this chapter, we recall that Aristotle's *Categories* is generally considered as the first part of the *Organon* ("tool"). The *Categories*, in the state we know it, is in two discernible parts. The first deals with the concept of category. To categorize originally meant "to say something of another thing" or "to bring an accusation against someone."

It may be recalled that Aristotelian terms like substance, relation, quality and quantity were not in philosophical usage but were taken from common usage. Therefore insofar as some content words start life as tool words we have an additional justification for general and wide use of tool words, even when our objective is to study the semantics of the text.

The second part of the *Categories* begins with forms of opposition, and ends with meanings of the verb "to have." Aristotle's authorship is not questioned, but the more prolix and less rigorous style raises questions about its place in relation to the *Organon*.

We used the e-text version available through Project Gutenberg [4]. It is divided into 3 sections. Within sections, there are parts, of which there are respectively 6, 2 and 7 for the three sections. Within parts, we used paragraphs, and, within paragraphs, sentences. We allowed for semi-colons to

385 dream reports (B. Sanders = –, o; Alta = |; remainder = *)

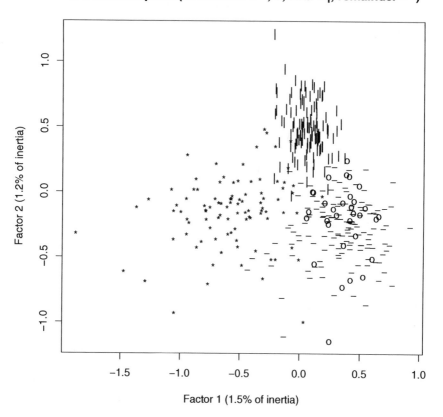

FIGURE 5.8

Principal factor plane of 385 dream reports, crossed by 584 content-oriented words.

Pride & Prejudice = 1; Persuasion = 2; Sense & Sensibility = 3

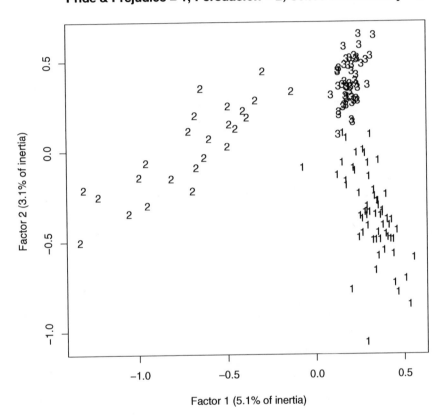

FIGURE 5.9
Principal factor plane of 135 chapters of Jane Austen novels, crossed by 745 content-oriented words.

be additional sentence terminators. A list of part headings follows (see Table 5.3). Overall this document consisted of 14,483 words. The number of unique words was found to be 1260. Examples from the top- and bottom-ranked: the 948, of 684, is 622, to 398, that 358, in 351, and 324, it 252, be 240, not 218, ... whose 1, wider 1, withstand 1, worse 1, wrestlers 1, wrestling 1, writing 1, yellow 1, yesterday 1, you 1. As in our other work, no grouping was undertaken in regard to singular and plural, since our assumption is that these can quite reasonably point in different semantic directions. Word forms were taken exactly as found, subject to deletion of punctuation. Given the ascii nature of the text used, there were no accented characters.

For a user-selected word file, cross-tabulation tables (contingency tables, correspondence tables) were determined at text resolution level 1, corresponding to sections; at level 2, corresponding to parts; at level 3, corresponding to paragraphs; and at level 4, corresponding to sentences.

Figure 5.10 shows an initial display, based on a set of 30 most commonly found words. We expect paragraphs to be located as centers of gravity of the sentences, and so on, and this can be observed modulo the fact that the representation shown is a two-dimensional projection of positions.

5.11.2   Structure of Presentation

The subdivisions of Aristotle's text are in the form of a succession of: 3 sections; spanning these, 15 parts; in turn, spanning these, 141 paragraphs; and in all 701 sentences. We chose the 15 parts as a suitable entry point for further investigation.

From the 1260 unique words, firstly we imposed the requirement that the word be present in the document 7 times or more. Extremely frequently occurring words ("the," "and," etc.) can be included, but such words cloud the output displays and make interpretation difficult. Therefore we removed pronouns, verbs and parts of verbs, and most adjectives. Quantifiers like "all," "some," and "each," were left. Most nouns were left. Following this manual selection process, we had 166 words.

The correspondence analysis of the 15 parts $\times$ 166 words data table led to the first and second factors explaining 13.7% and 12.8% of the inertia.

We begin with a short investigation of the word "grammar," and find in Figure 5.11 that it is closest to terms like "cases," "instances," "term," "word," "speech." It is less associated with {"terms," "derived"}, "attribute," "statement," "statements," {"case," "sense," "forms"}, "forms." Closeness here of course is linkage in the sense of use in nearby sentences. It is also linkage through use in the same context of a given part in the document.

We next looked at the use of words like "category," "class" and "genus": see Figure 5.12.

The word "category" is well-linked to words such as "quantity," "quality," "number," "degree."

TABLE 5.3

Heading levels in Aristotle's *Categories*.

Section I.

Part 1 – Homonyms, Synonyms and Derivatives

Part 2 – Simple and Composite Expressions

Part 3 – Concerning Predicates

Part 4 – The Eight Categories of the Objects of Thought

Part 5 – Substance

Part 6 – Quantity

Section II.

Part 7 – Relation

Part 8 – Qualities

Section III.

Part 9 – Action and Affection of the Other Categories Described

Part 10 – Four Classes of Opposites

Part 11 – Contraries Further Discussed

Part 12 – Uses of the Term "Prior"

Part 13 – Uses of the Term "Simultaneous"

Part 14 – Six Kinds of Motion

Part 15 – The Meanings of the Term "To Have"

Sections = S1, S2, S3; parts = numeric; paragraphs = +; sentences = .

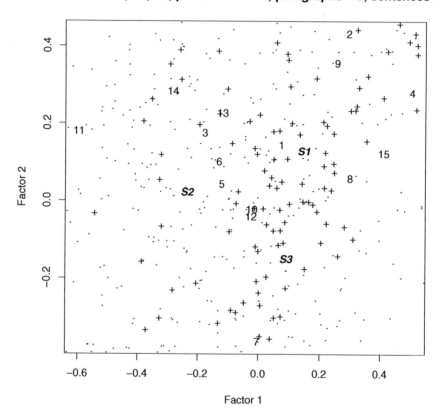

Factor 1

FIGURE 5.10

Principal factor plane with sections S1, S2 and S3; parts indicated by sequence numbers 1 to 15; paragraphs represented by "+"; and finally sentences represented by a dot.

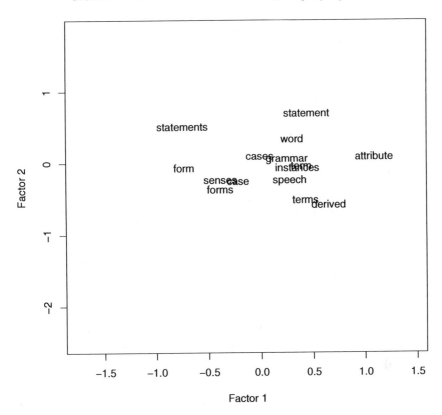

'Grammar' and fourteen other terms, displaying relationships

FIGURE 5.11
Principal factor plane of the word "grammar" and other, potentially related, terms, showing the relationships between them. The locations of these words are in the context of the correspondence analysis using all 166 words. Other words are not shown here.

The word "genus" is also well-linked to words like "species," "individual," "substance," "substances."

The word "class" though is not as well-linked. It is more free-standing, somewhere between "genus" and "category." In order to check that important words were not inadvertently omitted, Figure 5.13 shows the locations of all words. (Some of the locations that are not annotated are as follows: to the left are "socrates," "opposites," "opposite"; located towards but not on the far left are "necessary," "either"; towards the bottom near "distinct" and "sorts" are "motion," "rest," "alteration.")

The general evolution of the discussion in Aristotle's book is shown in Figure 5.14. Superimposed on the selected words is the set of parts, with sequence numbers 1 through 15. It is seen that "category" is highly associated with parts (4), 6, 8, 9 and (15), where those in brackets are less important. "Genus" is mostly covered in part 3, although parts 2 and 5 are also of some importance.

5.11.3   Evolution of Presentation

We studied the set of conjoined sections, parts, paragraphs and sentences, leading to the consideration of all 860 of these, crossed by 166 words selected as described in the previous section. Allowing for absence of any one of these words (which, after all, are a small subset of the total number of words) led to: no change in the sections and parts; 4 fewer paragraphs; and loss of a few sentences. In all, the 860 collections of words, represented by sections, parts, paragraphs and sentences, decreased to 821.

Correspondence analysis was carried out on the 166 words crossed by 821 collections of words. The first five percentages of inertia explained by the factors are: 3.15%, 2.92%, 2.46%, 2.27%, 2.10%. In Figure 5.15, we show the following:

1. All 166 words are located in the principal factor plane, denoted by asterisks.

2. Sections 1 to 3 are shown, with a line indicating the sequential evolution.

3. Parts 1 to 15 are shown with their sequence number.

Near "genus" is section 1, and parts 1 through 6. Near "category" is section 2, and parts 7, 8 and 9, and also 15.

We can go further and look at the evolution of word usage, as the argument evolves in the course of the book. The first of three listings below employs terms related to elucidation and discussion. The second listing below advances into technical and more directed discussion, using terminology and concepts developed. Finally the third listing below entails more enhanced and advanced application of the terminology and concepts now developed.

In total 32 words, including 'category,' 'class' and 'genus'

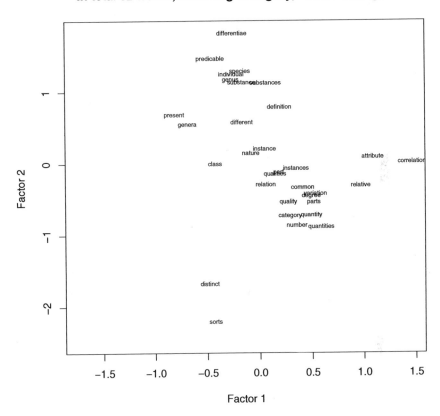

FIGURE 5.12
Principal factor plane displaying the words "category," "class," "genus" and in total 32 words, following a correspondence analysis using all 166 words.

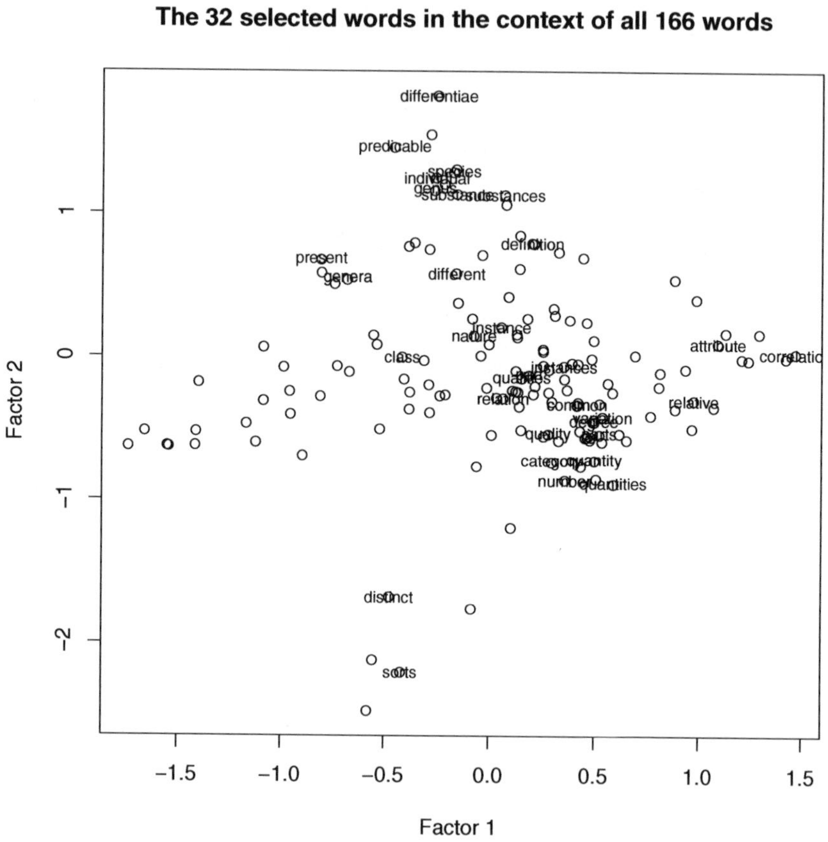

The 32 selected words in the context of all 166 words

FIGURE 5.13

Principal factor plane of the words "category," "class," "genus" and others among the 32 words selected, with the locations of all 166 words used for the correspondence analysis.

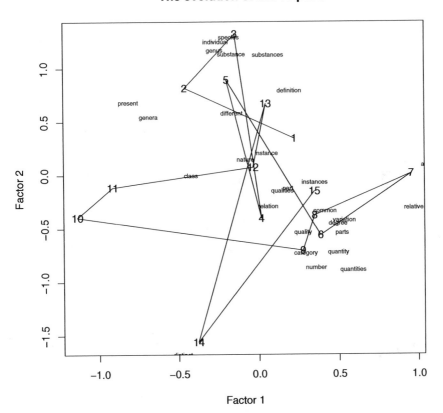

FIGURE 5.14
Principal factor plane with words displayed such as "category," "class,"
"genus" and others among the 32 words selected, and with the evolution
of the 15 parts located.

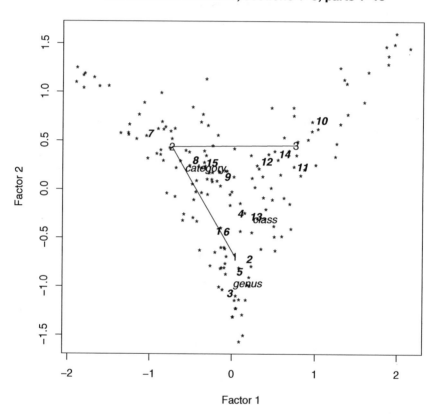

166 words as asterisks, sections 1–3, parts 1–15

FIGURE 5.15

Locations of words, sections and parts, in the principal plane.

Sentences in Section 1, relating to "category", also dealt with the following words:

"action"	"all"	"any"	"called"	"category"
"cubits"	"degree"	"double"	"everything"	"instance"
"kind"	"man"	"name"	"nature"	"nothing"
"number"	"object"	"place"	"position"	"predicated"
"properly"	"quality"	"quantitative"	"quantities"	"quantity"
"regard"	"relation"	"secondary"	"sense"	"some"
"something"	"sort"	"statement"	"substance"	"terms"
"thing"	"things"	"time"	"variation"	"way"
"white"				

Sentences in Section 2, relating to "category", also dealt with the following words:

"affections"	"affective"	"all"	"any"	"anything"
"applicable"	"body"	"called"	"cases"	"category"
"class"	"cold"	"contraries"	"contrary"	"dispositions"
"external"	"fact"	"forms"	"genera"	"genus"
"grammar"	"heat"	"individual"	"instance"	"instances"
"justice"	"knowledge"	"many"	"mark"	"mean"
"nothing"	"particular"	"place"	"predicated"	"properly"
"qualities"	"quality"	"quantity"	"reference"	"relation"
"relative"	"some"	"something"	"sort"	"terms"
"thing"	"things"	"virtue"	"white"	"whiteness"

Sentences in Section 3, relating to "category", also dealt with the following words:

"being"	"category"	"correlatives"	"corresponding"
"derived"	"double"	"knowledge"	"opposite"
"opposites"	"place"	"position"	"reference"
"relation"	"relative"	"rest"	"sense"
"some"	"something"	"term"	"terms"
"thing"	"things"	"time"	

5.12   Conclusion on Text Analysis Case Studies

Our aim in this chapter has been less with textual forensics and more with semantics. Clearly the interpretation of interdependencies between terms is facilitated by such analysis. The synchronic and diachronic study of textual structure is also made accessible by the displays resulting from factor projections.

Zipf's law relates term frequency of occurrence to numbers of terms. Our main concern with extremely frequent terms ("and," "the," etc.) is that such terms will cloud the interpretation. Showing far too much in the factor projection displays is not helpful for drawing conclusions. For this reason, in some of the discussion in this chapter, we used selected words only for display, even though the analysis was carried out on a far greater set of words.

Our main concern with very infrequent terms is that such terms will spoil the interpretation. Due to very limited co-occurrence, such terms are unusual, and can therefore overly influence the definition of the factors. We may find a factor with a dominating influence exerted from a rare term. The implication therefore is that rare terms should be excluded from the analysis.

6

Concluding Remarks

The following are the particularly unique aspects of this work:

- We have described a rich tradition of data analysis, presenting its historical linkages, underpinning theory and wide practical applicability.

- Software has been written in the graphical statistical R language, and in the programming language Java, together with text processing code in C, to permit quick hands-on experimentation, and indeed building this code into larger, customized software systems. This code is available on the web at address www.correspondances.info .

- We have focused on data coding, which constitutes an early and crucial step in the analysis. Data coding is also of direct relevance in such fields as machine learning, neural networks, statistics and exploratory data analysis.

- We have devoted much attention to the analysis of textual data, the domain of information retrieval. We have shown in many studies how correspondence analysis permits access to the content of textual information.

In the Foreword, in looking towards the future of multivariate data analysis, Jean-Paul Benzécri points to how natural science, and leading subareas of physics in particular, generate vast quantities of data with the sole aim of finding rare anomalous phenomena in this data.

To fully realise this ambitious goal of data analysis, the tools to be deployed may have to go far beyond what is described in this book, in the following ways.

- The selection of data is very important – as we have seen – in correspondence analysis. The use of supplementary elements leads straight away to an interactive analysis. The joint use of dimensionality reduction through factor analysis, and clustering, raises the number of combinatorial possibilities for the analysis still further.

- The end result of such a perspective is that the analyst is in conversation, in a dialog, with his or her data. The interactive environment supports this dialog. So, improved graphics are needed, and innovative human-machine interaction, in order to support such interactive work.

- A goal of the software becomes support for interactivity – responsive and dynamic analysis.

- A new software environment is needed for support of analysis that evolves in time. Therefore the aspects of interactive and dynamic support are both required for analysis.

References

[1] Adames, G. Foie gras et tradition: analyse des résultats d'un concours. *Les Cahiers de l'Analyse des Données*, XVIII, 389–398, 1993.

[2] Alkayar, A.M. Structure des introns au voisinage d'un point limite entre exon et intron sur une séquence d'ADN. *Les Cahiers de l'Analyse des Données*, XXI, 149–164, 1966.

[3] Anderson, E. The irises of Gaspe peninsula. *Bulletin of the American Iris Society*, 59, 2–5, 1935.

[4] Aristotle. *The Categories*, 350 BC. Translated by E.M. Edghill. Project Gutenberg e-text, www.gutenberg.net

[5] Aussem, A., Murtagh, F. and Sarazin, M. Dynamical recurrent neural networks – towards environmental time series prediction. *International Journal of Neural Systems*, 6, 145–170, 1995.

[6] Aussem, A., Murtagh, F. and Sarazin, M. Fuzzy astronomical seeing nowcasts with a dynamical and recurrent connectionist network. *International Journal of Neurocomputing*, 13, 353–367, 1996.

[7] Austen, J. *Sense and Sensibility*, 1811. Online version
http://www.pemberley.com/etext/SandS/index.html

[8] Austen, J. *Pride and Prejudice*, 1813. Online version
http://www.pemberley.com/janeinfo/pridprej.html

[9] Austen, J. *Persuasion*, 1817. Online version
http://www.pemberley.com/etext/Persuasion/index.html

[10] Aviation Accident Database and Synopses, National Transport Safety Board, 2003. http://www.landings.com/evird.acgi$pass*59062640!_h-www.landings .com/_landings/pages/search/rep-ntsb.html

[11] Bacon, F. *The New Organon.* (*Novum Organon*, 1620.) Cambridge University Press, Cambridge, 2000.

[12] Bastin, Ch., Benzécri, J.P., Bourgarit, Ch. and Cazes, P. *Pratique de l'Analyse des Données, Tome 2*, Dunod, 1980.

[13] Béhrakis, T. and Nicolacopulos, I. Analyse des réponses de 2000 électeurs à un thermomètre de sympathie vis-à-vis de personnalités politiques grecques. *Les Cahiers de l'Analyse des Données*, XIII, 233–238, 1988.

[14] Benzécri, F. Introduction à l'analyse des correspondances d'après l'analyse du commerce mondial des phosphates. *Les Cahiers de l'Analyse des Données*, X, 145–190, 1985.

[15] Benzécri, J.-P. *La Taxinomie*, 2nd ed., Dunod, 1976.

[16] Benzécri, J.-P. *L'Analyse des Données. II. Correspondances*, 2nd ed., Dunod, 1976.

[17] Benzécri, J.-P. and Benzécri, F. *Pratique de l'Analyse des Données, Vol. 1: Analyse des Correspondances. Exposé Élémentaire*, Dunod, 1980.

[18] Benzécri, J.-P. *Analyse des Données en Linguistique et Lexicologie*, Dunod, Paris, 1981.

[19] Benzécri, J.-P. *Histoire et Préhistoire de l'Analyse des Données*, Dunod, 1982.

[20] Benzécri, J.-P. Essai d'analyse des notes attribuées par un ensemble de sujets aux mots d'une liste. *Les Cahiers de l'Analyse des Données*, XIV, 73–98, 1989.

[21] Benzécri, J.-P. and Benzécri, F. Le codage linéaire par morceaux: réalisation et applications. *Les Cahiers de l'Analyse des Données*, XIV, 203–210, 1989.

[22] Benzécri, J.-P. and Benzécri, F. Codage linéaire par morceaux et équation personnelle. *Les Cahiers de l'Analyse des Données*, XIV, 331–336, 1989.

[23] Benzécri, J.-P. (translation T.K. Gopalan). *Correspondence Analysis Handbook*, Marcel Dekker, 1992.

[24] Benzécri, J.-P., Benzécri, F. et collab. *Pratique de l'Analyse des Données en Médecine Pharmacologie, Physiologie Clinique*, Statmatic, 1992.

[25] Benzécri, J.-P. Validité des échelles d'évaluation en psychologie et en psychiatrie et corrélations psychosociales. *Les Cahiers de l'Analyse des Données*, XVII, 55–86, 1992.

[26] Benzécri, J.-P. and Gopalan, T.K. Sur l'application des méthodes multidimensionnelles à une anthologie de données. (1) Morphométrie. *Les Cahiers de l'Analyse des Données*, XVIII, 427–446, 1993.

[27] Benzécri, J.-P. and Murtagh, F. Discrimination des jonctions entre exon et intron dans les séquences d'acide désoxyribonucéique. *Les Cahiers de l'Analyse des Données*, XXI, 133–148, 1996.

[28] Benzécri, J.-P. and Benzécri, F. Sources de programmes d'analyse de données en langage Pascal. (III) Élaboration de tableaux divers. (IIIG) Codage d'échelles suivant l'équation personnelle. *Les Cahiers de l'Analyse des Données*, XXII, 375–410, 1997/1998.

[29] Benzécri, J.-P. In memoriam: P. Bourdieu, 2002.

[30] Cailliez, F. and Pagès, J.-P. *Introduction à l'Analyse des Données.* SMASH (Société de Mathématiques Appliquées et de Sciences Humaines), Paris, 1976.

[31] Capaccioli, M., Ortolani, S. and Piotto, G. Empirical correlations between globular cluster parameters and mass function morphology. *Astronomy and Astrophysics*, 244, 298–302, 1991.

[32] Chaieb, S. Variation de la concentration plasmatique d'une substance au cours d'une perfusion et après celle-ci: cas du dinitrate d'isosorbide. *Les Cahiers de l'Analyse des Données*, IX, 43–57, 1984.

[33] Chomsky, N. *Syntactic Structures.* Mouton, The Hague, 1957.

[34] Coombs, C.H. *A Theory of Data.* Wiley, New York, 1964.

[35] Deerwester, S., Dumais, S., Furnas, G., Landauer, T. and Harshman, R. Indexing by latent semantic analysis. *Journal of the American Society for Information Science*, 41, 391–407, 1990.

[36] Domhoff, G.W. *The Scientific Study of Dreams: Neural Networks, Cognitive Development and Content Analysis*, American Psychological Association, 2003.

[37] Doob, J.L. *Stochastic Processes*, Wiley, 1953.

[38] DreamBank. Repository of dream reports, www.dreambank.net, 2004.

[39] Easterly, W., Kremer, M., Pritchett, L. and Summers, L.H. Good policy or good luck? Country growth performance and temporary shocks. *Journal of Monetary Economics*, 32, 459–83, 1993.

[40] Economic Growth Research – Easterly, Kremer, Pritchett, Summers Data Set, http://www.worldbank.org/research/growth/ddekps.htm, World Bank.

[41] Esmieu, D., Gopalan, T.K. and Maiti, G.D. Sur l'utilisation des échelles numériques dans les études de marché préparant l'introduction d'un produit nouveau. *Les Cahiers de l'Analyse des Données*, XVIII, 399–426, 1993.

[42] Fisher, R.A. The use of multiple measurements in taxonomic problems. *The Annals of Eugenics*, 7, 179–188, 1936.

[43] Gallego, F.J. Codage flou en analyse des correspondances. *Les Cahiers de l'Analyse des Données*, VII, 413–430, 1982.

[44] Greenacre, M.J. *Theory and Applications of Correspondence Analysis*, Academic Press, 1984.

[45] Greenacre, M.J. Interpreting multiple correspondence analysis. *Applied Stochastic Models and Data Analysis*, 7, 195–210, 1991.

[46] Honkela, T., Pulkki, V. and Kohonen, T. Contextual relations of words in Grimm tales, analyzed by self-organizing map. In F. Fogelman-Soulié and P. Gallinari, Eds., *Proc. International Conference on Artificial Neural Networks, ICANN-95*, EC2 et Cie, Paris, 1995, pp. 3–7.

[47] Jambu, M. *Classification Automatique pour l'Analyse des Données. 1. Méthodes et Algorithmes*, Dunod, 1978.

[48] Kohonen, T. *Self-Organizing Maps*, Springer-Verlag, 1995.

[49] Krzanowski, W.J. Attribute selection in correspondence analysis of incidence matrices. *Applied Statistics*, 42, 529–541, 1993.

[50] Lebart, L., Morineau, A. and Warwick, K.M. *Multivariate Descriptive Statistical Analysis*, Wiley, 1984.

[51] Lebart, L., Salem, A. and Berry, L. Recent developments in the statistical processing of textual data. *Applied Stochastic Models and Data Analysis*, 7, 47–62, 1991.

[52] Le Roux, B. and Rouanet, H. *Geometric Data Analysis*, Springer, 2004.

[53] Loslever, P., Guerra, T.M. and Roger, D. Analyse des questionnaires en ergonomie: L'appréciation des réglages d'un poste de travail. *Les Cahiers de l'Analyse des Données*, XIII, 175–196, 1988.

[54] Maïza, S. Le commerce mondial des phosphates de 1973 à 1980. *Les Cahiers de l'Analyse des Données*, IX, 7–32, 1984.

[55] Manly, B.F.J. *Multivariate Statistical Methods: A Primer*. Chapman and Hall, 1985.

[56] Mantegna, R.N. and Stanley, H.E. *An Introduction to Econophysics*, Cambridge University Press, 2000.

[57] McGibbon Taylor, B., Leduc, P. and Tibeiro, J.S. Analyse des réponses des étudiants à un questionnaire relatif au mémoire de recherche de la maîtrise en administration des affaires. *Les Cahiers de l'Analyse des Données*, XIV, 337–346, 1989.

[58] Milligan, G.W. and Cooper, M.C. A study of standardization of variables in cluster analysis. *Journal of Classification*, 5, 181–205, 1988.

[59] Miyamoto, S. *Fuzzy Sets in Information Retrieval and Cluster Analysis*, Kluwer, 1990.

[60] Murtagh, F. A survey of recent advances in hierarchical clustering algorithms. *The Computer Journal*, 26, 354–359, 1983.

[61] Murtagh, F. Structure of hierarchic clusterings: implications for information retrieval and for multivariate data analysis. *Information Processing and Management*, 20, 611–617, 1984.

[62] Murtagh, F. *Multidimensional Clustering Algorithms*, CompStat Lectures Volume 4, Physica-Verlag, 1985.

[63] Murtagh, F. and Heck, A. *Multivariate Data Analysis*, Kluwer, 1987.

[64] Murtagh, F. and Sarazin, M. Nowcasting astronomical seeing: a study of ESO La Silla and Paranal. *Publications of the Astronomical Society of the Pacific*, 105, 932–939, 1993.

[65] Murtagh, F. Neural networks and related massively parallel methods in statistics: an overview, *International Statistical Review*, 62, 275–288, 1994.

[66] Murtagh, F., Aussem, A. and Sarazin, M. Nowcasting astronomical seeing: towards an operational approach. *Publications of the Astronomical Society of the Pacific*, 107, 702–707, 1995.

[67] Murtagh, F. and Hernández-Pajares, M. The Kohonen self-organizing map method: an assessment. *Journal of Classification*, 12, 165–190, 1995.

[68] Nishisato, S. *Analysis of Categorical Data: Dual Scaling and Its Applications*, University of Toronto Press, 1980.

[69] Noël-Jorand, M.C., Reinert, M., Giudicelli, S. and Drassa, D. Discourse analysis in psychosis: characteristics of hebephrenic subject's speech. Proc. JADT 2000: 5^{ièmes} Journées d'Analyse Statistique des Données Textuelles.

[70] Ockerbloom, J.M. Grimm's Fairy Tales, 2003. http://www-2.cs.cmu.edu/~spok/grimmtmp

[71] Pack, P. and Jolliffe, I.T. Influence in correspondence analysis. *Applied Statistics*, 41, 365–380, 1992.

[72] Poinçot, P., Lesteven, S. and Murtagh, F. A spatial user interface to the astronomical literature. *Astronomy and Astrophysics Supplement Series*, 130, 183–191, 1998.

[73] Poinçot, P., Lesteven, S. and Murtagh, F. Maps of information spaces: assessments from astronomy. *Journal of the American Society for Information Science*, 51, 1081–1089, 2000.

[74] Prechelt, L. Proben1: A set of neural network benchmark problems and benchmarking rules. Technical Report, Fakultät für Informatik, Universität Karlsruhe, D-76128 Karlsruhe, Germany. ftp://ftp.ira.uka.de/pub/neuron/proben1.tar.gz, 1994.

[75] Ross, R.M. *An Elementary Introduction to Mathematical Finance: Options and Other Topics*, 2nd ed., Cambridge University Press, 2002.

[76] Rouanet, H. and Le Roux, B. *Analyse des Données Multidimensionnelles*, Dunod, 1993.

[77] Sadaka, M. Enquête sur les marques d'arak au Liban. *Les Cahiers de l'Analyse des Données*, XIX, 387–394, 1994.

[78] Samuelson, P.A. Proof that properly anticipated prices fluctuate randomly. *Industrial Management Review*, 6, 41–45, 1965.

[79] Schneider, A. and Domhoff, G.W. The quantitative study of dreams. http://dreamresearch.net, 2004.

[80] Sneath, P.H.A. and Sokal, R.R. *Numerical Taxomomy*. Freeman, 1973.

[81] Torgerson, W.S. *Theory and Methods of Scaling*. Wiley, New York, 1958.

[82] Towell, G.G. and Shavlik, J.W. Interpretation of artificial neural networks: mapping knowledge-based neural systems into rules. In *Advances in Neural Information Processing Systems*, Vol. IV, Morgan Kaufman, 1992.

[83] Unicode 16-bit character code standard, www.unicode.org.

[84] Velleman, P.F. and Wilkinson, L. Nominal, ordinal, interval and ratio typologies are misleading. *The American Statistician*, 47, 65–72, 1993.

[85] M. Volle. *Analyse des Données*, 2nd ed., Economica, 1981.

[86] WebSOM. WEBSOM – Self-Organizing Map for Internet Exploration, various papers, interactive software system, http://websom.hut.fi, 1997.

[87] Hongyuan Zha, Xiaofeng He, Ding, C., Simon, H. and Ming Gu, Bipartite graph partitioning and data clustering. *Proc. ACM 10th International Conference on Information and Knowledge Management – CIKM 2001*, Nov. 5–10, Atlanta, GA, pp. 25–31, 2001.

Index